PROBABILISTIC METHODS
IN APPLIED MATHEMATICS

Volume 2

CONTRIBUTORS TO THIS VOLUME

A. T. BHARUCHA-REID
STANLEY GUDDER
W. M. WONHAM

PROBABILISTIC METHODS
IN APPLIED MATHEMATICS

Edited by $\begin{bmatrix} A. & T. & Bharucha\text{-}Reid \end{bmatrix}$

Albert

CENTER FOR RESEARCH IN PROBABILITY
DEPARTMENT OF MATHEMATICS
WAYNE STATE UNIVERSITY
DETROIT, MICHIGAN

VOLUME 2

ACADEMIC PRESS New York and London 1970

ACADEMIC PRESS, INC.
111 Fifth Avenue, New York, New York 10003

United Kingdom Edition published by
ACADEMIC PRESS, INC. (LONDON) LTD.
Berkeley Square House, London W1X 6BA

LIBRARY OF CONGRESS CATALOG CARD NUMBER: 68–18657

PRINTED IN THE UNITED STATES OF AMERICA

LIST OF CONTRIBUTORS

A. T. Bharucha-Reid, Center for Research in Probability, Department of Mathematics, Wayne State University, Detroit, Michigan

Stanley Gudder, Department of Mathematics, University of Denver, Denver, Colorado

W. M. Wonham,* Center for Dynamical Systems, Division of Applied Mathematics, Brown University, Providence, Rhode Island and NASA/Electronics Research Center, Cambridge, Massachusetts

* Present address: Department of Electrical Engineering, University of Toronto, Toronto 5, Canada.

PREFACE

An examination of the current literature of science, engineering, and technology shows that modern probability theory is exerting a profound influence on the formulation of mathematical models and on the development of theory in many applied fields; in turn, problems posed in applied fields are motivating research in probability theory. In addition, probability theory is finding increased applications in, and interaction with, other branches of mathematics. This interaction is very desirable, and of particular interest are the uses of probabilistic methods in various branches of analysis and those developments in abstract probability theory which are of value in the field of applied mathematics.

As is well known, progress in any science is highly dependent upon developments in methodology. Applied mathematics is a good example of this phenomenon. Within recent years research in stochastic processes, functional analysis, and numerical analysis has led to the development of powerful methodological tools for the applied mathematician; and there is considerable evidence that applied mathematicians are indeed using the results of research in the above fields to formulate theories and study more realistic mathematical representations of concrete natural phenomena.

This present serial publication, which will be published in several volumes at irregular intervals, is devoted to the role of modern probability theory, in particular the theory of stochastic processes, in the general field of applied mathematics. We will not attempt to define "applied mathematics," but will assume that its objective is the development and utilization of mathematical methods to understand natural phenomena and technological systems quantitatively. We propose to cover a rather wide range of general topics, special topics, and problem areas in the mathematical sciences, and each volume of this serial publication will contain several articles, each being written by an expert in his field. Although each article will be reasonably self-contained and fully referenced, the reader is assumed to be familiar with measure-theoretic probability and the basic classes of stochastic processes, and the elements of functional analysis. The individual articles are not intended to be popular expositions of the survey type, but are to be regarded, in a sense, as brief monographs which can serve as intro-ductions to specialized study and research.

In view of the above aims, the nature of the subject matter, and the manner in which the text is organized, these volumes will be addressed to a broad audience of mathematicians specializing in probability theory and its applications, applied mathematicians working in those areas in which probabilistic methods are being employed, physicists, engineers, and other scientists interested in probabilistic methodology and its potential applicability in their respective fields.

January, 1970 A. T. BHARUCHA-REID

CONTENTS

RANDOM DIFFERENTIAL EQUATIONS IN CONTROL THEORY

W. M. Wonham

CONTENTS OF VOLUME 1

RANDOM ALGEBRAIC EQUATIONS

A. T. Bharucha-Reid

CENTER FOR RESEARCH IN PROBABILITY
DEPARTMENT OF MATHEMATICS
WAYNE STATE UNIVERSITY
DETROIT, MICHIGAN

I. Introduction

Let
$$F_n(z) = a_0 + a_1 z + \cdots + a_n z^n, \tag{1.1}$$
where $z = x + iy$, $x, y \in R$, be a polynomial of degree n. The coefficients a_i, $i = 0, 1, \ldots, n$, are given real or complex numbers with $a_n \neq 0$. The relation $\xi = F_n(z)$ defines a mapping of the complex plane Z into itself. On equating the polynomial to zero we get the equation
$$F_n(z) = 0, \tag{1.2}$$
which is called an *algebraic equation*. Any value of z satisfying Eq. (1.2) is called a *root* or *zero* of the polynomial $F_n(z)$; and the problem of solving Eq. (1.2) is that of determining the *solution set* $S = \{z : F_n(z) = 0\}$. As is well known, the *fundamental theorem of algebra* asserts that there exists at least one complex number, z_0 say, such that $F_n(z_0) = 0$. Hence S is a nonempty subset of Z, and contains in the case $a_n \neq 0$ at most n points.

Our purpose in this article is to survey research on several problems associated with certain stochastic or probabilistic analogs of Eq. (1.2). Since a polynomial is uniquely determined by its $n + 1$ coefficients $(a_0, a_1, \ldots, a_n)$, probabilistic analogs of Eq. (1.2) are obtained by considering the coefficients to be real (or complex)-valued random variables.

Although the study of random algebraic equations is of independent theoretical interest (in that their study leads to probabilistic generalizations of classical results concerning polynomials), many applied problems in mathematical physics, engineering, and statistics lead to the study of algebraic equations whose coefficients are random variables. An algebraic equation with random coefficients will arise if, for example, the coefficients are subject to random error. This might occur when the coefficients of a polynomial are computed from experimental data, or when numerical procedures require that truncated values of the computed coefficients be used.

Random algebraic equations also arise in the spectral theory of matrices with random elements (cf. [10]), with subsequent applications in mathematical physics (cf. Mehta [44]; Porter [49]) and multivariate statistical analysis (cf. Anderson [1]; Wilks [61]). Consider, for example, the 2×2 random matrix
$$A(\omega) = \begin{pmatrix} a_{11}(\omega) & a_{12}(\omega) \\ a_{21}(\omega) & a_{22}(\omega) \end{pmatrix}.$$

In this case the associated determinantal equation $|A(\omega) - \lambda I| = 0$ is a random quadratic equation, and, as in the deterministic case, it will have two solutions $\lambda_1(\omega)$ and $\lambda_2(\omega)$. In the probabilistic case, however, it is the joint distribution of $\lambda_1(\omega)$ and $\lambda_2(\omega)$ as a function of the distribution of the random elements $a_{ij}(\omega)$ of $A(\omega)$ which is of interest; the main problem is to compute the measures or probabilities of the three possible solution sets (i.e., $\{\omega: \lambda_1(\omega), \lambda_2(\omega)$ real and unequal$\}$, $\{\omega: \lambda_1(\omega), \lambda_2(\omega)$ real and equal$\}$, and $\{\omega: \lambda_1(\omega), \lambda_2(\omega)$ complex conjugates$\}$).

In addition to the role of random algebraic equations in the spectral theory of random matrices, there are several important points of contact between the theories of random algebraic and random operator equations. As is well known, the solution of a homogeneous ordinary differential (or difference) equation of nth order with constant co-efficients is expressed as a function of the roots of the characteristic polynomial associated with the differential (or difference) operator. However, in the case of a random differential (or difference) operator, for example,

$$L(\omega)[y(x)] = \sum_{k=0}^{n} a_k(\omega)(d^k y/dx^k), \qquad (1.3)$$

or

$$L(\omega)[y_x] = \sum_{k=0}^{n} a_k(\omega)\tau^k[y_x], \qquad (1.4)$$

where τ denotes the translation operator

$$\tau^k[y_x] = y_{x+k}, \qquad k = 0, 1, 2, \ldots, n,$$

it is impossible to obtain explicit expressions for the roots of the associated *random characteristic polynomials*

$$F_n(\lambda, \omega) = \sum_{k=0}^{n} a_k(\omega)\lambda^k. \qquad (1.5)$$

in terms of the coefficients of the operator and to determine the distri-bution of the solution of the random equation.

The remainder of this article is divided into five sections. In Section II we present a general discussion of random algebraic polynomials and some of their properties. Sections III and IV are devoted to the funda-mental problems associated with random algebraic equations, namely the determination of the *number, and expected number, of roots* of a random algebraic polynomial and the *distribution* of the roots. In Section V we consider some limit theorems associated with the number of roots of a random algebraic polynomial; finally, in Section VI we discuss some connections between random matrices and random algebraic equations.

II. Random Algebraic Polynomials

A. INTRODUCTION

In this section we introduce the concept of a random algebraic polynomial and discuss certain properties of random polynomials which will be used in subsequent sections of this article. We first introduce a few definitions.

Definition 2.1. Let $(\Omega, \mathscr{A}, \mu)$ be a fixed probability space, and let $\{a_k(\omega)\}_{k=0}^{\infty}$ be a sequence of complex-valued random variables defined on Ω. Then, the formal power series

$$F(z, \omega) = \sum_{k=0}^{\infty} a_k(\omega)z^k, \qquad (2.1)$$

where $z \in Z$, $\omega \in \Omega$, is called a *random power series*, and the random variables $a_k(\omega)$ are called the *coefficients of the random power series*.

For every fixed $z \in Z$, a random power series is a sum of complex-valued random variables, and is a complex-valued random variable if the series converges (in some appropriate sense). Hence, if for fixed $z_0 \in Z$ the power series converges, then

$$F(z_0, \omega) = \sum_{k=0}^{\infty} a_k(\omega)z_0{}^k \qquad (2.2)$$

is an $\mathscr{A}$-measurable mapping of Ω into Z. If we now fix $\omega \in \Omega$, the random power series is an ordinary (or deterministic) power series; and, we can speak of the power series

$$F(z, \omega_0) = \sum_{k=0}^{\infty} a_k(\omega_0)z^k, \qquad (2.3)$$

as a *realization* (or *sample function*) of the random power series (2.1).

The study of random power series was initiated by Borel and Steinhaus. At the present time there is a rather large amount of literature dealing with probabilistic methods in classical analysis. The theory of random power series and related topics are active areas of research. We refer to the lecture notes of Arnold [4] and the beautiful book of Kahane [37].

We now consider the specialization of random power series to random polynomials. Let

$$\Omega_0 = \{\omega\colon a_k(\omega) \neq 0 \text{ finitely often}\} \subset \Omega.$$

Definition 2.2. A random power series with $\mu(\Omega_0) = 1$ is called a *random polynomial*. $d(\omega) = \max\{k\colon a_k(\omega) \neq 0\}$ is a real-valued random variable with $d(\omega) < \infty$ almost surely, where $d(\omega)$ is called the *degree* of the random polynomial. If $d^* = \min\{n\colon d(\omega) < n \text{ almost surely}\} < \infty$, then d^* is called the *highest degree* of the random polynomial.

It is of interest to determine conditions for which a random power series is a random polynomial. We state the following result due to Arnold [2]:

Theorem 2.1. *A random power series is a random polynomial if and only if*

$$\mu\left(\lim_{n} \inf\{\omega\colon a_n(\omega) = 0\}\right) = 1.$$

We omit the proof, since it follows easily from the definition of the set Ω_0, and the definition of the limit inferior of a sequence of sets.

In view of the above, we can say that a *random algebraic polynomial of degree n* ($n \geq 1$) is a polynomial of the form

$$F_n(z, \omega) = a_0(\omega) + a_1(\omega)z + \cdots + a_n(\omega)z^n, \tag{2.4}$$

where $z \in Z$, and the coefficients $a_0(\omega)$, $a_1(\omega)$, ..., $a_n(\omega)$ are complex-valued random variables; hence they are of the form $a_k(\omega) = \alpha_k(\omega) + i\beta_k(\omega)$, where $\alpha_k(\omega)$ and $\beta_k(\omega)$ are real-valued random variables. To say that $F_n(z, \omega)$ is of degree n, means that for the infinite-tuples $(a_0(\omega), a_1(\omega), \ldots)$ we have $a_k(\omega) = 0$ almost surely for all $k \geq n + 1$. Hence $d(\omega) \leq n$ almost surely; and since $\mu(\{\omega\colon a_n(\omega) \neq 0\}) > 0$ we have $d^* = n$.

Let D be a domain of the complex plane Z. A random function $f(z, \omega)$ is said to be a *random analytic function* if almost all of its realizations can be analytically continued in the region D (cf., Belyaev [81]). A random polynomial $F_n(z, \omega)$ is an example of a random analytic function. Also, given a random function $f(z, \omega)$, $z \in D$, with $f(z, \omega) \in L_2(\Omega)$ for all $z \in D$, we can talk about the random function being *analytic in quadratic mean* in D; that is, there exists a function $f'(z, \omega) \in L_2(\Omega)$ such that

$$\lim_{h \to 0} \left\| \frac{f(z + h, \omega) - f(z, \omega)}{h} - f'(z, \omega) \right\|_2 = 0.$$

The above will obtain if and only if the covariance function $K(z_1, z_2) = \mathscr{E}\{f(z_1, \omega)\overline{f(z_2, \omega)}\}$ defined on $D \times D$ is an analytic function of the pair (z_1, z_2) in $D \times D$.

Kac [36] has considered another "model" for random polynomials which can be described as follows: Given a deterministic polynomial of degree n with real coefficients, say $F_n(x)$, we associate the point $\mathbf{a} = (a_0, a_1, \ldots, a_n)$, and restrict our attention to the sphere

$$S_{n+1}(1) = \left\{ \mathbf{a} \colon \|\mathbf{a}\| = \left(\sum_{k=0}^{n} |a_k|^2 \right)^{1/2} = 1 \right\};$$

that is, the unit sphere in Euclidean space E_{n+1}. In this case we take $\Omega = S_{n+1}(1)$, and the probability measure μ is the normalized Lebesgue measure of the surface of $S_{n+1}(1)$.

Random polynomials can also be considered within the framework of probabilistic functional analysis [10]. Let $f_n(x)$, $x \in [a, b]$ be a polynomial of degree n with real coefficients; and let P denote the class of all polynomials $f_n(x)$. Then, as is well known, P is a subalgebra of the space of continuous functions $C[a, b]$. We can now consider a random polynomial F_n as an $\mathscr{A}$-measurable mapping

$$F_n \colon (\Omega, \mathscr{A}) \to (P, \mathscr{B}_0),$$

where $\mathscr{B}_0 = P \cap \mathscr{B}$, and $\mathscr{B}$ denotes the σ-algebra of Borel subsets of $C[a, b]$. That is F_n is a *generalized random variable* with values in the subalgebra P of the Banach space $C[a, b]$.

As in the case of deterministic polynomials, the roots of a random polynomial $F_n(z, \omega)$ form the solution set of the *random algebraic equation*

$$F_n(z, \omega) = a_0(\omega) + a_1(\omega)z + \cdots + a_n(\omega)z^n = 0. \qquad (2.5)$$

In the case of random algebraic equations the solution set will be a random subset of the complex plane Z, i.e., the solution set S will depend on $\omega \in \Omega$. There are two main problems in the study of random algebraic equation: (I) the determination of the number, and expected number, of roots of $F_n(z, \omega)$ which are in a specified subset of Z, and (II) the determination of the distribution of the roots, or, equivalently, the probability measure of the solution set. Problems I and II will be discussed in Sections III and IV, respectively, however, in order to carry out a rigorous probabilistic analysis of random algebraic equations it is necessary to first consider several measurability questions, namely (1) the measurability of the roots of $F_n(z, \omega)$, and (2) the measurability of the number of roots, say $N(B, \omega)$, of $F_n(z, \omega)$ in $B \subset Z$. The next two subsections of this section are devoted to these questions.

B. Measurability of the Roots

Let $F_n(z, \omega)$, where $z = x + iy \in Z$ and $\omega \in \Omega$, be a random polynomial of degree n with coefficients $a_k(\omega) = \alpha_k(\omega) + i\beta_k(\omega)$. Now $F_n(z, \omega)$ will have n roots, say $\xi_1, \xi_2, \ldots, \xi_n$, and it is clear that these roots will be functions of ω. However, it is not obvious that the $\xi_k(\omega)$, $k = 1, 2, \ldots, n$ will be random variables; hence one of the first problems to be considered is that of the measurability of the roots $\xi_k(\omega)$. The theorem we state and prove below, which is due to Hammersley [31], asserts that $\xi_k = (x_k, y_k)$, for each $k = 1, 2, \ldots, n$, is a Borel measurable function of $\mathbf{a}(\omega) = (\alpha_0, \alpha_1, \ldots, \alpha_n, \beta_0, \beta_1, \ldots, \beta_n)$.

We will need the following lemma:

Lemma 2.1. *There exists a fixed indexing* $z_1, z_2, \ldots, z_m$ *of any given fixed set of m distinct points in Z such that*

$$|z_1 - e^{i\pi/h}| > |z_2 - e^{i\pi/h}| > \cdots > |z_m - e^{i\pi/h}| > 0$$

for all sufficiently large positive integers h, say $h \geq \Phi(z_1, z_2, \ldots, z_m)$, *where* Φ *is some function of* $z_1, z_2, \ldots, z_m$.

Theorem 2.2. *Consider the random polynomial*

$$F_n(z, \omega) = \sum_{k=0}^{n} a_k(\omega)z^k, \qquad where \ a_k(\omega) = \alpha_k(\omega) + i\beta_k(\omega)$$

and $z = x + iy$. *Then, there exists an indexing* $\xi_1, \xi_2, \ldots, \xi_n$ *of the roots of* $F_n(z, \omega)$ *such that* $\xi_k = (x_k, y_k)$ *is a Borel measurable function of* $\mathbf{a}(\omega)$ *for each* $k = 1, 2, \ldots, n$.

Proof. We first assume that $\mathbf{a}(\omega)$ is fixed, and that $a_n(\omega) = 1$. In this case the roots of $F_n(z, \omega)$ are fixed and all finite. We next assume that there are m distinct roots among the n roots, hence we can use the indexing of Lemma 2.1 to denote these roots $\xi_1, \xi_2, \ldots, \xi_m$. Now the roots are uniquely determined by $\mathbf{a}$. Therefore the function Φ of the lemma is a function of $\mathbf{a}$, say $\Phi(\xi_1, \xi_2, \ldots, \xi_m) = \Phi(\mathbf{a})$.

For each integer $h = 1, 2, \ldots$, define numbers γ_{jh} by means of the following identity in z:

$$\sum_{j=0}^{n} a_j(z + e^{i\pi/h})^j = \sum_{j=0}^{n} \gamma_{jh} z^j. \tag{2.6}$$

For every fixed $t = 1, 2, \ldots, n$ let φ_t be an arbitrary real number such that $\varphi_t \in [0, 1]$; and for $h, t = 1, 2, \ldots$ define $X_{h,t}$ as a function of $\boldsymbol{\varphi} = (\varphi_1, \varphi_2, \ldots, \varphi_n)$ by the recurrence relations

$$X_{h,t} = \begin{cases} \varphi_t, & \text{for} \quad t = 1, 2, \ldots, n \\ -\displaystyle\sum_{j=0}^{n} \gamma_{jh} X_{h,t-n+j}, & \text{for} \quad t = n+1, n+2, \ldots. \end{cases} \tag{2.7}$$

Now, put

$$Y_{h,t} = \begin{cases} X_{h,t+1} X_{h,t}, & \text{for} \quad X_{h,t} \neq 0 \\ 0, & \text{for} \quad X_{h,t} = 0; \end{cases} \tag{2.8}$$

and put

$$f(\mathbf{a}) = \lim_{h \to \infty} \int_0^1 \int_0^1 \cdots \int_0^1 \lim_{t \to \infty} (Y_{h,t} + e^{i\pi/h})\, d\varphi_1\, d\varphi_2 \cdots d\varphi_n. \tag{2.9}$$

We now show that the above limit exists; and that $f(\mathbf{a})$ is a Borel measurable function of $\mathbf{a}$ (subject to the condition $a_n = 1$), and $f(\mathbf{a}) = \xi_1$.

From (2.6) we see that the condition $a_n = 1$ implies $\gamma_{nh} = 1$. Therefore, the linear difference equation (2.7) has a solution of the form

$$X_{h,t} = \sum_{k=1}^{m} P_k(t)(\xi_k - e^{i\pi/h})^t, \qquad t = 1, 2, \ldots, \tag{2.10}$$

since the distinct roots of the polynomial on the right-hand-side of (2.6) are $z = \xi_k - e^{i\pi/h}$, $k = 1, 2, \ldots, m$. The functions $P_k(t)$ in (2.10) are polynomials in t whose coefficients are functions of $\boldsymbol{\varphi}$. We remark that the polynomials have degree less than n and are all identically zero if and only if $\boldsymbol{\varphi} = \mathbf{0}$. For $\boldsymbol{\varphi} \neq \mathbf{0}$, let k_0 denote the smallest value of k such that $P_k(t) \neq 0$. Now, consider any fixed value of h such that $h \geq \Phi(\mathbf{a})$. Utilizing Lemma 2.1, we have

$$|\xi_{k_0} - e^{i\pi/h}| > |\xi_{k_0+1} - e^{i\pi/h}| > \cdots > |\xi_m - e^{i\pi/h}| > 0. \tag{2.11}$$

Hence, from (2.10) and (2.8), we have

$$\lim_{t \to \infty} Y_{h,t} = \begin{cases} \xi_{k_0} - e^{i\pi/h}, & \text{for} \quad \boldsymbol{\varphi} \neq \mathbf{0} \\ 0, & \text{for} \quad \boldsymbol{\varphi} = \mathbf{0}. \end{cases} \tag{2.12}$$

Now k_0 is clearly a function of $\boldsymbol{\varphi}$; however we shall show that $k_0 = 1$ for almost all $\boldsymbol{\varphi}$.

For $t = 1, 2, \ldots, n$, it follows from (2.7) and (2.10) that

$$\varphi_t = \sum_{k=1}^{m} P_k(t)(\xi_k - e^{i\pi/h})^t. \tag{2.13}$$

Concerning the coefficients of the polynomials $P_k(t)$, there are a total of n coefficients in all of the $P_k(t)$ taken together. Now, if any of these coefficients are zero, then it follows from (2.13) that there is a linear relation between $\varphi_1, \varphi_2, \ldots, \varphi_n$; that is, $\boldsymbol{\varphi}$ lies on a hyperplane. Hence, except perhaps when $\boldsymbol{\varphi}$ lies on one or more of n hyperplanes in the unit hypercube, none of the $P_k(t)$ is zero. Therefore, from (2.12) and (2.9), for $h \geq \Phi(\mathbf{a})$

$$\int_0^1 \int_0^1 \cdots \int_0^1 \lim_{t \to \infty} (Y_{h,t} + e^{i\pi/h}) \, d\varphi_1 \, d\varphi_2 \cdots d\varphi_n$$
$$= (\xi_{k_0} - e^{i\pi/h}) + e^{i\pi/h} = \xi_{k_0} = \xi_1, \tag{2.14}$$

since k_0 is the smallest value of k such that $P_k(k) \neq 0$. Finally, from (2.9), $f(\mathbf{a})$ exists and $f(\mathbf{a}) = \xi_1$ for every fixed $\mathbf{a}$ subject to the condition $a_n = 1$.

The above result was obtained for every fixed $\mathbf{a}$, i.e., for every fixed $\omega \in \Omega$. Now consider the coefficient vector $\mathbf{a}$ as a function of ω, with $a_n(\omega) = 1$ almost surely. From (2.6), it is clear that the coefficients γ_{jh} are Borel measurable functions of $\mathbf{a}(\omega)$ and h; hence, by (2.7) and (2.8), $Y_{h,t}$ is a Borel measurable function of $\mathbf{a}(\omega)$, h, and $\boldsymbol{\varphi}$. The Borel measurability of $f(\mathbf{a}(\omega)) = \xi_1(\omega)$ follows from the fact that the limit and integral of Borel measurable functions are Borel measurable.

We now drop the condition $a_n(\omega) = 1$ almost surely. If $a_n(\omega) \neq 0$ almost surely, then we can divide all of the coefficients of $F_n(z, \omega)$ by $a_n(\omega)$ without affecting the root $\xi_1(\omega)$. If $a_n(\omega) = 0$ almost surely, then we define $\xi_1(\omega) = (+\infty, +\infty)$. Hence it follows that, in general, $\xi_1(\omega) = (x_1(\omega), y_1(\omega))$ is a Borel measurable function of $\mathbf{a}(\omega)$.

In order to obtain the next root, the above process can be applied to the polynomial

$$\sum_{k=0}^{n-1} \tilde{a}_k(\omega) z^k,$$

whose roots are those of $F_n(z, \omega)$ excepting the one root at $z = \xi_1$. The coefficients $\tilde{a}_k(\omega)$ are defined as follows:

(i)   For $a_n(\omega) \neq 0$ almost surely,

$$\tilde{a}_k(\omega) = \begin{cases} 0, & \text{for} \quad k = n \\ a_{k+1}(\omega) - \xi_1(\omega)\tilde{a}_{k+1}(\omega), & \text{for} \quad k = 0, 1, \ldots, n-1; \end{cases}$$

and

(ii)   For $a_n(\omega) = 0$ almost surely

$$\tilde{a}_k(\omega) = a_k(\omega), \qquad k = 0, 1, \ldots, n-1.$$

Since a Borel measurable function of a Borel measurable function is Borel measurable, it follows from the above that the $\tilde{a}_k(\omega)$ are Borel measurable functions of $\mathbf{a}(\omega)$. Therefore, the next root $\xi_2(\omega)$ of $F_n(z, \omega)$ is a Borel measurable function of $\mathbf{a}(\omega)$.

Repeated application of the above procedure provides the required indexing of the roots, and establishes the Borel measurability of the roots $F_n(z, \omega)$.

C. Measurability of the Number of Roots

Let $F_n(z, \omega)$ be a random polynomial of degree n defined on the domain D of the complex plane Z; and let $\mathscr{F}(D)$ be the σ-algebra of Borel subsets of D. Let $N_n(B, \omega)$ denote of *number of roots* of $F_n(z, \omega)$ in $B \subset D$. In this subsection we show that $N_n(B, \omega)$ is a random variable for every $B \in \mathscr{F}(D)$, and that $\mathscr{E}\{N_n(B, \omega)\}$, the *expectation of the number of roots*, is a measure on $\mathscr{F}(D)$.

The following result is due to Arnold [4].

Theorem 2.3. *Let $F_n(z, \omega)$, $z \in D$, be a random polynomial of degree n; and let $\mathcal{N} = \{B: B \subset D, N_n(B, \omega) \text{ is a random variable}\}$. Then $\mathscr{F}(D) \subset \mathcal{N}$.*

Proof. We want to prove the measurability of the mapping $N_n(B, \omega)$: $\Omega \to \{0, 1, \ldots, n\}$. This will be done by showing that $N_n(B, \omega)$ can be represented as the composition of three measurable mappings; hence $N_n(B, \omega)$ itself will be measurable.

In Section II.A we pointed out that a random polynomial $F_n(z, \omega)$, $z \in D$, could be considered as a generalized random variable with values in the algebra of polynomials; that is, $F_n(z, \omega)$ is an $\mathscr{A}$-measurable mapping $F_n: (\Omega, \mathscr{A}) \to (P, \mathscr{B}_0)$, where P is the subalgebra of polynomials of the space of bounded continuous functions $C(D)$, and $\mathscr{B}_0 = P \cap \mathscr{B}$, and $\mathscr{B}$ is the σ-aglebra of Borel subsets of $C(D)$. We remark that $\mathscr{B}_0$ is also the minimal σ-algebra generated by the open sets of the compact open topology τ_c in P.

Now let τ_v be the vague topology[1] in the set M of all Borel measures on $(D, \mathscr{F}(D))$. Consider the topological spaces (P, τ_c) and (M, τ_v). Then the mapping $\tilde{N} : (P, \tau_c) \to (M, \tau_v)$, where $\tilde{N}(F_n)$ denotes the number of roots of F_n, is continuous; and since $\tilde{N}$ is continuous it is a $\mathscr{F}_0$-measurable mapping of P into $(M, \mathscr{B}(\tau_v))$, where $\mathscr{B}(\tau_v)$ is the σ-algebra of Borel sets of (M, τ_v).

Now let φ be a positive measurable function on $(D, \mathscr{F}(D))$; and consider the mapping $T : (M, \mathscr{B}(\tau_v)) \to (\{0, 1, \ldots, n\}, \mathscr{B}\{0, 1, \ldots, n\})$ defined by $T(v) = \int \varphi \, dv$. T is a measurable mapping since the class of functions for which T is measurable is monotone, and contains the functions $f \in C(D)$ according to the definition of the vague topology. In particular, we can take $\varphi = \chi_B$ (the indicator function of $B \in \mathscr{F}(D)$), so that

$$\int \varphi \, dv = v(B) = T_B(v).$$

It follows from the above that

$$N_n(B, \omega) = T_B(\tilde{N}(F_n(z, \omega))), \tag{2.15}$$

which is measurable.

We remark that by taking $D = Z$, which we can always do in the case of polynomials, the measurability of $N_n(B, \omega)$ can be established in a very simple and direct fashion. Firstly, we observe that the mapping $N_n(B, \omega)$ is measurable if and only if $\{\omega : N_n(B, \omega) = k\} \in \mathscr{A}$ for $k = 0, 1, \ldots, n$. Secondly, we know from Theorem 2.2 that there exists an indexing such that the roots $\xi_1(\omega), \ldots, \xi_n(\omega)$ of $F_n(z, \omega)$ are measurable. Finally,

$$\{\omega : N_n(B, \omega) = k\}$$
$$= \{\omega : \text{exactly } k \text{ of the roots}$$
$$\xi_1(\omega), \ldots, \xi_n(\omega) \text{ are in } B\} \in \mathscr{A}.$$

---

[1] Let S denote a separable topological Hausdorff space, and let $C(S)$ denote the set of all continuous real-valued functions on S with compact support. A sequence of measures $v_1, v_2, \ldots$ on S is said to *converge vaguely* to the measure v if $\int_S g(s) v_n \, (ds) \to \int_S g(s) v \, (ds)$ for every $f \in C(S)$ (cf. Grenander [29, pp. 35–36]). The vague topology τ_v is the topology of vague convergence.

We also have the following corollary.

Corollary 2.1. *A random algebraic polynomial* $F_n(z, \omega)$, $z \in D$, *defines by its number of roots* N_n *a random point distribution in* D; *that is*

   (i) $N_n: \mathscr{F}(D) \times \Omega \to \{0, 1, \ldots, n\}$,
   (ii) $N_n(B, \omega)$ *is a measure on* $\mathscr{F}(D)$ *for almost all* $\omega \in \Omega$,
   (iii) $N_n(B, \omega)$ *is measurable on* $(\Omega, \mathscr{A}, \mu)$ *for all* $B \in \mathscr{F}(D)$,
   (iv) $N_n(B, \omega) \le n$ *almost surely for every relatively compact set* $B \subset D$.

Since $\{N_n(B, \omega), B \in \mathscr{F}(D)\}$ is a random point distribution, it follows that the *expectation of the number of roots*

$$\mathscr{E}\{N_n(B, \omega)\} = v_n(B) \tag{2.16}$$

is a *measure* on $\mathscr{F}(D)$.

III. The Number of Roots of a Random Algebraic Polynomial

A. Introduction

Let

$$F_n(z, \omega) = a_0(\omega) + a_1(\omega)z + \cdots + a_n(\omega)z^n \tag{3.1}$$

be a random algebraic polynomial of degree n, and let $N(B, \omega)$ denote the number of roots of $F_n(z, \omega)$ (equivalently, the number of solutions of the random algebraic equation $F_n(z, \omega) = 0$) in $B \subset D$, where $D \subset Z$ is the domain of $F_n(z, \omega)$. During the past 30–40 years most of the studies on random algebraic polynomials have been concerned with the estimation of $N_n(B, \omega)$ and $v_n(B) = \mathscr{E}\{N_n(B, \omega)\}$ when various assumptions are made concerning the coefficients $a_0(\omega)$, $a_1(\omega)$, ..., $a_n(\omega)$. In this introductory section we present a brief summary of some of the early studies on the estimation of $N_n(B, \omega)$ and $v_n(B)$; and in Sections III.B–D we present a discussion of some recent studies, the results of these studies being presented in detail in order to illustrate the various analytical techniques that are utilized in the study of the number of roots of random algebraic polynomials. Although we will restrict our discussion to random algebraic polynomials, the techniques and many of the results are applicable to *random trigonometric polynomials* since, for example, the trigonometric polynomial

$$T_n(\theta, \omega) = \sum_{k=0}^{n} a_k(\omega) \cos k\theta \qquad (3.2)$$

is a random algebraic polynomial of degree n in $\cos \theta$. We refer to the papers of Das [14] and Dunnage [15] for results on the number and average number of real roots of random trigonometric polynomials. In a paper on the roots of deterministic trigonometric polynomials, Zinger [62] has obtained some interesting results on the number of roots of even and odd trigonometric polynomials in terms of the properties of the roots of the associated algebraic polynomials. An investigation of the roots of algebraic and trigonometric polynomials utilizing probabilistic generalizations of Zinger's theorems is indicated since we feel that such an investigation might lead to new methods of studying random polynomials.

To turn now to a summary of early results, and recently obtained related results, on the number of roots of random algebraic polynomials, Bloch and Pólya [12], in one of the first studies on random polynomials, considered the equation $F_n(\omega, x) = 0$, $x \in R$, when (i) $a_0(\omega) = 1$ a.s., and (ii) $\mathscr{P}\{a_k(\omega) = 1\} = \mathscr{P}\{a_k(\omega) = -1\} = \mathscr{P}\{a_k(\omega) = 0\} = \frac{1}{3}$. With the above assumptions on the coefficients, they showed that

$$v_n(R) = \mathscr{E}\{N_n(R, \omega)\} = O(n^{1/2}) \qquad (n \to \infty).$$

Littlewood and Offord [38, 39] in a series of papers which initiated the general study of random algebraic equations considered the equation $F_n(\omega, x) = 0$, $x \in R$, under the assumption that the coefficients $a_k(\omega)$ are independent and identically distributed real-valued random variables. Three cases were considered, namely:

(i) The coefficients are normal (or Gaussian) random variables with mean 0 and standard deviation 1.
(ii) The coefficients are uniformly distributed in $(-1, 1)$.
(iii) The coefficients $a_k(\omega)$, $k = 1, 2, \ldots, n$, assume the values $+1$ or -1 with equal probability, and $a_0(\omega) = 1$ a.s.

The main results of Littlewood and Offord are as follows:

Theorem 3.1. *In each of the three cases listed above, and for[2] $n \geq n_0$,*

$$\mathscr{P}\{N_n(R, \omega) > 25(\log n)^2\} \leq (12(\log n)/n).$$

[2] In particular, the theorem is true for $n_0 = 2000$.

Corollary 3.1. *For $n \geq n_0$,*

$$v_n(R) \leq 25(\log n)^2 + 12 \log n.$$

Theorem 3.2. *In each of the three cases listed above,*

$$\mathscr{P}\{N_n(R, \omega) < \alpha[\log n/(\log \log n)^2]\} < A/\log n,$$

where α and A are absolute constants.

Hence, the above results established lower and upper bounds for the number of real roots $N_n(R, \omega)$, as well as an upper bound for the expected number of real roots $v_n(R)$, in each of the three cases considered. Littlewood and Offord [40] also considered the following case:

(iv) The coefficients $a_k(\omega)$ are independent random variables, with $a_0(\omega) = c_0$ a.s., and $\mathscr{P}\{a_k(\omega) = c_k\} = \mathscr{P}\{a_k(\omega) = -c_k\} = \frac{1}{2}$ $(k = 1, 2, \ldots, n)$, where the c_k $(k = 0, 1, \ldots, n)$ are fixed complex numbers.

In this case they obtained the following result:

Theorem 3.3.

$$\mathscr{P}\left\{N_n(R, \omega) \geq 10 \log n\left(\log \frac{\sum_{k=0}^{n} |c_k|}{|c_0 c_n|^{1/2}} + 2(\log n)^5\right)\right\}$$
$$\leq A \log \log n/\log n,$$

where A is an absolute constant.

We remark that the estimates provided by Theorems 3.1–3.3 are in some ways rather crude; since Theorems 3.1 and 3.3 are liable to count some complex roots along with the real roots, and Theorem 3.2 ignores all real roots except those in a small neighborhood of $x = 1$.

Kac [34, 35] (cf. also [36]) improved the results of Littlewood and Offord by showing that in the case of a random algebraic equation of degree n with the coefficients $a_k(\omega)$ independent and normally distributed with mean 0 and standard deviation 1

$$v_n(R) = \frac{4}{\pi} \int_0^1 \frac{[1 - \Phi_n(x)]^{1/2}}{1 - x^2} \, dx, \tag{3.3}$$

$$\Phi_n(x) = (n + 1)x^n \left[\frac{1 - x^2}{1 - x^{2n+2}}\right] \tag{3.4}$$

for $x \neq \pm 1$. A measure-theoretic statement of the above result is as follows: v_n is m-continuous on $\mathscr{B}(R)$, where m is Lebesgue measure on $\mathscr{B}(R)$; that is there exists a m-density ρ_n of v_n such that

$$v_n(B) = \int_B \rho_n(x)\, dx \qquad (3.5)$$

for all $B \in \mathscr{B}(R)$, where

$$\rho_n(x) = \begin{cases} \dfrac{1}{\pi} \dfrac{[1 - \Phi_n^{\,2}(x)]^{1/2}}{|1 - x^2|}, & x \neq \pm 1 \\[3mm] \dfrac{1}{\pi}\left[\dfrac{n(n+2)}{12}\right]^{1/2}, & x = \pm 1. \end{cases} \qquad (3.6)$$

From the above it follows that

$$\lim_{n \to \infty} \rho_n(x) = \frac{1}{\pi}\frac{1}{|1 - x^2|}, \qquad x \neq \pm 1, \qquad (3.7)$$

and

$$\rho_n(1) = \rho_n(-1) = \frac{1}{\pi}\left[\frac{n^2 - 1}{12}\right]^{1/2} \sim \frac{n}{\pi(12)^{1/2}}, \qquad (3.8)$$

The graph of $\rho_n(x)$ is given in Fig. 1, and shows that the real roots tend on the average to concentrate around $x = \pm 1$. Equation (3.8) indicates how pronounced this tendency is.

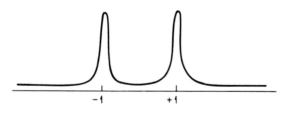

$$-1 \qquad\qquad +1$$

FIG. 1.

Kac also obtained from (3.4) the asymptotic result

$$v_n(R) \sim \frac{2}{\pi} \log n \qquad (n \to \infty), \qquad (3.9)$$

and the estimate

$$v_n(R) \leq \frac{2}{\pi} \log n + \frac{14}{\pi}, \qquad n \geq 2. \qquad (3.10)$$

Calculations show that on the average there are very few real roots; for example, for $n = 10^3$ we have $v_n(R) \leq 9.2$, and for $n = 10^6$ we have $v_n(R) \leq 14$. The asymptotic result (3.4) is also valid if the coefficients $a_k(\omega)$ are uniformly distributed in $[-1, 1]$.

Erdős and Offord [19] also showed that (3.9) is valid for the discrete case, namely the case when each coefficient $a_k(\omega)$ is $+1$ or -1 with equal probability, and the $a_k(\omega)$ are independent. In particular, they proved the following result:

Theorem 3.4. *The number of real roots of most of the polynomials* $F_n(x, \omega)$ *is*

$$\frac{2}{\pi} \log n + o((\log n)^{2/3} \log(\log n));$$

and the measure of the exceptional set does not exceed $o((\log \log n)^{-1/3})$.

Theorem 3.4 can also be formulated as follows:

$$\mathscr{P}\left\{ N_n(R, \omega) \neq \frac{2}{\pi} \log n + o((\log n)^{2/3} \log(\log n)) \right\}$$

$$\leq o((\log \log n)^{1/3}).$$

At this point, it is of interest to note the connection between the results of Rice [53] on the zeros of real-valued random functions (with subsequent applications in the theory of random noise) and the results of Kac. Let $g(x, \omega)$, $x \in R$, be a real-valued random function, and let $f(s, t; x)$ denote the joint probability density for $g(x, \omega) = s$ and $g'(x, \omega) = t$. Rice gave a heuristic proof of the following result: *The expected number of zero crossings of* $g(x, \omega)$ *is m-continuous with density*

$$\rho(x) = \int_{-\infty}^{\infty} |t| \, f(0, t; x) \, dt. \tag{3.11}$$

If, as pointed out in Section II.A, we consider the random polynomial $F_n(x, \omega)$ as a real-valued random function, then applying (3.11) to the case considered by Kac we obtain the formula for $\rho_n(x)$ as given by (3.6). We refer to Cramér and Leadbetter [13] for a discussion of zero crossing problems for stationary random functions.

Stevens [58] improved Kac's estimate of $v(R)$ for Gaussian random polynomials showing that

$$\frac{2}{\pi} \log n - 0.6 < v_n(R) < \frac{2}{\pi} \log n + 1.4, \tag{3.12}$$

and proved [58, 59] the following general result:

Theorem 3.5. *Let $F_n(\omega, x) = 0$ be a random algebraic equation whose coefficients $a_k(\omega)$ are independent real-valued random variables with $\mathscr{E}\{a_k(\omega)\} = 0$, $\mathscr{E}\{a_k^2(\omega)\} = 1$, $\mathscr{E}\{a_k^4(\omega)\} < B$, and the probability densities of $a_0(\omega)$ and $a_n(\omega)$, say $f_0(x)$ and $f_n(x)$, are such that*

$$f_i(x) \le B/(1 + x^{1 + 1/B}) \qquad (i = 0, n)$$

for some finite B, then $v_n(R) \sim (2/\pi) \log n$ $(n \to \infty)$.

Two other recent studies of interest are those of Evans [22] and Dunnage [16]. Evans proved the following two theorems for Gaussian random polynomials which can be regarded as "strong" versions of Theorems 3.1 and 3.2, due to Littlewood and Offord.

Theorem 3.6. *There exist an integer n_0 and a set $\Omega_0 \subset \Omega$ with $\mu(\Omega_0) \le A/[\log n_0 - \log \log \log n_0]$ such that, for each $n > n_0$ and all $\omega \in \Omega - \Omega_0$, $N_n(R, \omega) \le \alpha(\log \log n)^2 \log n$, where α and A are constants.*

Theorem 3.7. *There exist an integer n_0 and a set $\Omega_0 \subset \Omega$ with $\mu(\Omega_0) \le B \log \log n_0/\log n_0$ such that, for each $n > n_0$ and all $\omega \in \Omega - \Omega_0$, $N_n(R, \omega) \ge \beta \log n/\log \log n$, where β and B are constants.*

The above results are "strong" in the following sense: Theorem 3.1 is of the form

$$\mathscr{P}\left\{\frac{N_n(R, \omega)}{v_n(R)} < \alpha\right\} \to 1, \qquad n \to \infty,$$

while Theorem 3.6 is of the form

$$\mathscr{P}\left\{\sup_{n > n_0} \frac{N_n(R, \omega)}{v_n(R)} < \alpha\right\} \to 1, \qquad n \to \infty.$$

Dunnage has given more general forms of Theorems 3.2 and 3.3 of Littlewood and Offord. His general form of Theorem 3.3, for polynomials with complex-valued random coefficients, will be discussed in Section III.D.

B. Real Roots of Cauchy Algebraic Polynomials

In this section, following Logan and Shepp [41], we evaluate $v_n((a, b))$ and give an asymptotic estimate of $v_n((a, b))$ for a *Cauchy algebraic polynomial*; that is a random algebraic polynomial

$$F_n(x, \omega) = \sum_{k=0}^{n} a_k(\omega)x^k, \qquad x \in (-\infty, \infty),$$

whose coefficients are independent real-valued random variables with a common Cauchy distribution. In order to obtain $v_n((a, b))$, the expected number of roots of a Cauchy algebraic polynomial in the interval (a, b), we will use the Kac–Rice formula as given by (3.5) and (3.11); that is

$$v_n((a, b)) = \int_a^b dx \int_{-\infty}^{\infty} |t| \, g(0, t; x) \, dt, \qquad (3.13)$$

where $g(s, t; x)$ is the joint probability density for $F_n(x, \omega) = s$ and $F_n'(x, \omega) = t$.

   In order to obtain $g(0, t; x)$ we use characteristic function methods. In the Cauchy case the characteristic function of the coefficients is given by

$$\mathscr{E}\{\exp\{i\lambda a_k(\omega)\}\} = e^{-|\lambda|}, \qquad k = 0, 1, \ldots, n \qquad (3.14)$$

An application of the Fourier inversion formula gives

$$g(s, t; x) = \frac{1}{(2\pi)^2} \int_{-\infty}^{\infty} \int_{-\infty}^{\infty} \exp\{-is\lambda - it\gamma\} \Phi(\lambda, \gamma) \, d\lambda \, d\gamma, \quad (3.15)$$

where

$$\Phi(\lambda, \gamma) = \mathscr{E}\{\exp\{iF_n(x, \omega)\lambda + iF_n'(x, \omega)\gamma\}\}$$

$$= \exp\left\{-\sum_{k=0}^{n} |x^k\lambda + kx^{k-1}\gamma|\right\}.$$

If we now put $\gamma = x\lambda$, $\lambda = -uv$, and $s = 0$; then

$$g(0, t; x) = \frac{2x}{(2\pi)^2} \, \text{Re}\left[\int_0^{\infty} \exp\{itxv\} \, dv \int_{-\infty}^{\infty} \exp\left\{-v \sum_{k=0}^{n} |u - k| x^k\right\} v \, du\right].$$

Integrating the above expression with respect to v, we obtain

$$g(0, t; x) = \frac{2x}{(2\pi)^2} \int_{-\infty}^{\infty} \text{Re}[A(u, x) + itx]^{-2} \, du,$$

where we put

$$A(u, x) = \sum_{k=0}^{n} |u - k| x^k.$$

If we now multiply $g(0, t; x)$ by $|t|$ and integrate with respect to t, we obtain

$$\int_{-\xi/x}^{\xi/x} |t| \, g(0, t; x) \, dt = \frac{1}{\pi^2 x} \int_{-\infty}^{\infty} \left(\frac{A - uA'}{A} \right) \left(\frac{\xi^2}{\xi^2 + A^2} \right) du.$$

Hence as $\xi \to \infty$

$$\int_{-\infty}^{\infty} |t| \, g(0, t; x) \, dt = \frac{1}{\pi^2 x} \left\{ \int_0^n \log A(u, x) \, du + n - n \log A(n, x) \right.$$

$$\left. + (M/N) n((nM - N)/M) \right\}. \tag{3.16}$$

In (3.16) we have put $M = \sum_{k=0}^{n} x^k$ and $N = \sum_{k=0}^{n} kx^k$. From (3.13) it follows that (3.16) represents the expected number of real roots at x. Hence

$$v_n((a, b)) = \mathscr{E}\{N_n((a, b), \omega)\} = \frac{1}{\pi^2} \int_a^b dx \int_0^n \log z_n(u, x) \, du, \tag{3.17}$$

where

$$z_n(u, x) = \sum_{k=0}^{n} |(u - k)x^k| \left/ \left| \sum_{k=0}^{n} (u - k)x^k \right| \right..$$

Logan and Shepp also showed that

$$v_n(R) \sim \alpha \log n \qquad (n \to \infty), \tag{3.18}$$

where

$$\alpha = \frac{8}{\pi^2} \int_0^{\infty} \frac{\xi e^{-\xi}}{\xi - 1 + 2e^{-\xi}} \, d\xi.$$

It is of interest to compare the expected number of real roots of Gaussian and Cauchy algebraic polynomials. If, for Cauchy algebraic polynomials, we put

$$\rho_n(x) = \frac{1}{\pi^2 x} \int_0^n \log z_n(u, x) \, du,$$

then

$$\rho_n(1) \sim \frac{n(\pi/2 - 1)}{\pi^2}.$$

For Gaussian algebraic polynomials, the result of Kac [cf. (3.8)] is

$$\rho_n(1) \sim \frac{n}{\pi\sqrt{12}}.$$

Hence we can conclude that *near* $x = 1$ there are *fewer* real roots in the Cauchy case.

On the other hand, Kac (3.9), and Stevens (Theorem 3.5) showed that $v_n(R) \sim (2/\pi) \log n \ (n \to \infty)$. If we compare $\alpha \approx 0.7413$ with $2/\pi \approx 0.6366$, we observe that although the order of growth is the same, there are *more* real roots in the Cauchy case.

Logan and Shepp [41a] have also considered the general case in which the coefficients $a_k(\omega)$ are independent random variables with common characteristic function $\exp\{-|\lambda|^\alpha\}$, often referred to as the symmetric stable law with index $\alpha, 0 < \alpha \le 2$. In this case they prove that $v_n(R) \sim c \log n \ (n \to \infty)$, where

$$c = c(\alpha) = [4/(\pi\alpha)^2] \int_{-\infty}^{\infty} dx \log[|x - y|^\alpha/|x - 1|^\alpha] e^{-y} \, dy.$$

Hence $c(2) = 2/\pi$ and $c(0^+) = 1$. For $\alpha = 2$ the result of Kac for the normal case is obtained; and for $\alpha = 1$ the result given above for the Cauchy case is obtained.

C. Real Roots of Random Algebraic Polynomials—General Case

In this section we consider in detail the problem of estimating the number of real roots of a random algebraic polynomial $F_n(x, \omega)$, $x \in R$, when the coefficients $a_k(\omega)$ are subject to the following conditions:

(i) The $a_k(\omega)$, $k = 0, 1, \ldots, n$ (where n is sufficiently large), are independent and identically distributed real-valued random variables; and

(ii) $\mathscr{E}\{a_k(\omega)\} = 0$; and $\sigma^2 = \mathscr{E}\{a_k^2(\omega)\}$, and $\mathscr{E}\{|a_k(\omega)|^2\}$ are finite and nonzero.

The two theorems which we will state and prove are due to Samal [56]. The proofs, which utilize Jensen's formula (cf. Rudin [54, pp. 299–302]) and some results of Offord [48]; Berry [9]; and Esseen [21] on sums of independent random variables, are modifications of those of Littlewood and Offord. Similar methods have been employed in recent studies (cf., for example, Dunnage [16]).

The first theorem we prove is as follows:

Theorem 3.8. *Let $F_n(\omega, x)$ be a random polynomial with coefficients subject to the conditions stated above. Then, for $n \geq n_0$*

$$N_n(R, \omega) \leq \alpha(\log n)^2,$$

where α is a positive absolute constant. The measure of the exceptional set tends to zero as n tends to infinity.[3]

The proof of the above theorem utilizes two lemmas about sums of independent random variables, due to Offord [48], which we state without proof.

Lemma 3.1. *Let $y_1, y_2, \ldots, y_n$ be a sequence of independent real-valued random variables, and let $\alpha_i = \mathscr{E}\{y_i\}$, $\beta_i = \mathscr{E}\{(y_i - \alpha_i)^2\}$, $\gamma_i^3 = \mathscr{E}\{|y_i - \alpha_i|^3\}$; and put $2\rho^{1/3} = \min_i(\alpha_i/\beta_i)$. Then for all $n > 1$*

$$\max_{-\infty < t < \infty} \mathscr{P}\left\{ t - x \leq \sum_{i=1}^{n} y_i \leq t + x \right\} \leq C \frac{\log n}{\rho^3 \sqrt{n}} \left(\log n + \frac{\rho x}{\min_i \alpha_i} \right),$$

where C is an absolute constant.

Lemma 3.2. *Let $y_1, y_2, \ldots, y_n$ be a sequence of independent real-valued random variables, and let α_i, β_i, and γ_i be defined as in Lemma 3.1. Let s denote any subsequence $i_1, i_2, \ldots, i_m$ out of the $2^n - n$ possible subsequences of at least two terms which can be formed from the natural numbers $1, 2, \ldots, n$. Put*

$$2\rho_s^{1/3} = \min_{i = i_1, \ldots, i_n} (\alpha_i/\beta_i), \qquad \beta(s) = \min_{i = i_1, \ldots, i_n} \beta_i.$$

Then

$$\max_{-\infty < t < \infty} \mathscr{P}\left\{ t - x \leq \sum_{i=1}^{n} y_i \leq t + x \right\}$$

$$\leq C \min_s \frac{\log m(s)}{\rho^3(s) m^{1/2}(s)} \left[\log m(s) + \frac{\rho(s)x}{\beta(s)} \right],$$

where C is an absolute constant, $m(s)$ is the number of terms in the subsequence s, and the minimum is taken for all the $2^n - n$ possible subsequences of s.

---

[3] Theorem 3.8 can also be formulated as follows: *There exists a constant $\alpha > 0$ such that*

$$\mathscr{P}\{N_n(R, \omega) > \alpha(\log n)^2\} \to 0 \qquad as \quad n \to \infty.$$

Proof of Theorem 3.8. We first remark that although $x \in (-\infty, \infty)$ it is sufficient to restrict our attention to the number of roots in the interval $(0, 1)$; since to each root of $F_n(x, \omega)$ in $(0, 1)$ there corresponds a root of $F_n(-x, \omega)$ in $(-1, 0)$, and conversely. Also, if $F_n(x, \omega)$ has a root in $(1, \infty)$, then $x^n F_n(\xi, \omega)$, where $\xi = 1/x$, has a root in $(0, 1)$. Therefore it is sufficient to consider the number of roots in the interval $(0, 1)$, for $N_n(R, \omega)$ and the measure of the exceptional set are each four times the corresponding estimates for the interval $(0, 1)$.

Since we are restricting our attention to the interval $(0, 1)$, we will first obtain an estimate of the number of roots in $[\frac{1}{2}, 1]$, and then an estimate of the number of roots in $(0, \frac{1}{2})$. Let p be a fixed number greater than $1/\log 2$, and let $k = [p \log n]$. We denote by Γ_m circles with centers $x_m = 1 - (\frac{1}{2})^m$ and with radii $r_m = \frac{1}{2}(1 - x_m)$, where $m = 1, 2, \ldots, k$, $p \log n$; and we put $r_0 = 1/n$ when $x_0 = 1$. It is clear that the union of the circles $\Gamma_0, \Gamma_1, \ldots, \Gamma_k, \Gamma_{p \log n}$ covers the interval $[\frac{1}{2}, 1]$. That the circle $\Gamma_{p \log n}$ extends beyond the circle Γ_0 follows from the inequality

$$r_{p \log n} + r - (x_0 - x_{p \log n}) > \frac{1}{2^{p \log n + 1}} > 0, \tag{3.19}$$

since $p > 1/\log 2$. Let Γ_m' denote the circle concentric with Γ_m and with radius $2r_m$. In view of the above, all circles Γ_m' are interior to $|z| = 1 + 2/n$.

Now, if $g(z)$ is a regular function, the number of roots of $g(z)$ in a circle with center z_0 and radius r, say $N(|z - z_0| < r)$, satisfies the inequality

$$N(|z - z_0| < r) \leq \log \frac{\max_{|z - z_0| < r'} |g(z)|}{|g(z_0)|} \left(\log \frac{r'}{r} \right)^{-1}, \tag{3.20}$$

where r', with $r' > r$, denotes the radius of a concentric circle. (This result is a consequence of the well-known formula of Jensen.) If we now put $z_0 = z_m$, $r' = 2r_m$, and $r = r_m$, (3.20) enables us to state that the number of roots of any random polynomial in Γ_m is at most

$$\log \frac{\max_{|z| \leq 1 + 2/n} |F_n(z, \omega)|}{|F_n(x_m, \omega)|} (\log 2)^{-1}. \tag{3.21}$$

An application of Chebychev's inequality yields

$$\mathscr{P}\{|a_k(\omega)| \geq (n + 1)\} < \sigma^2/(n + 1)^2;$$

hence

$$\mathscr{P}\{|a_k(\omega)| < (n + 1)\} > 1 - \sigma^2/(n + 1)^2, \tag{3.22}$$

for $0 \le k \le n + 1$. Therefore,

$$\mathscr{P}\left\{ \max_{|z| \le 1 + 2/n} |F_n(z, \omega)| \le (n + 1) \sum_{i=0}^{n} (1 + 2/n)^i \right\} \ge 1 - \sigma^2/(n + 1). \quad (3.23)$$

For $m > 0$ put

$$M(m) = \log n/(-\log(1 - \tfrac{1}{2}m)) = \log n/(-\log x_m), \quad (3.24)$$

and let $I(m)$ denote the integral part of $M(m)$. In order to apply Lemma 3.2 to $F_n(x_m, \omega)$, $m > 0$, we take the subsequence s to be the subsequence $1, 2, \ldots, I(m)$, and put

$$\alpha_i = 0, \quad \beta_i^2 = \sigma^2 x_m^{2i} \quad \gamma_i^3 = \tau^3 x_m^{3i}, \quad \text{where} \quad \tau^3 = \mathscr{E}\{|a_i(\omega)|^3\}.$$

Using (3.24), we can write

$$2\rho_s^{1/3} = \min_{i=1, 2, \ldots, I(m)} \left(\frac{\beta_i}{\gamma_i} \right) = 2\rho^{1/3}$$

(say), and

$$\beta(s) = \min_{i=1, 2, \ldots, I(m)} \sigma x_m^i = \sigma x_m^{I(m)} > \sigma x_m^{M(m)} = \sigma/n.$$

Applying Lemma 3.2 we obtain

$$\mathscr{P}\{|F_n(x_m, \omega)| \le 1/n\} = C \frac{\log I(m)}{\rho^3 [I(m)]^{1/2}} \left[\log I(m) + \frac{\rho}{\sigma} \right]$$

$$\le C \frac{\log M(m)}{\rho^3 [\tfrac{1}{2} M(m)]^{1/2}} \left[\log M(m) + \frac{\rho}{\sigma} \right], \quad (3.25)$$

since $\tfrac{1}{2}M(m) < I(m) \le M(m)$. If we now apply Lemma 3.1 to $F_n(x_0, \omega)$, we get

$$\mathscr{P}\{|F_n(x_0, \omega)| \le 1/n\} \le C \frac{\log(n + 1)}{\rho^3 [n + 1]^{1/2}} \left[\log(n + 1) + \frac{\rho}{\sigma} \right], \quad (3.26)$$

since $\min_i \beta_i = \sigma \min_i x_0^i = \sigma > \sigma/n$. Therefore, from (3.19), (3.22), (3.23), (3.25), and (3.26), it follows that except for a set of measure at most

$$\frac{\sigma^2}{n + 1} + C \frac{\log M(m)}{\rho^3 \tfrac{1}{2} M(m)} \left[\log M(m) + \frac{\rho}{\sigma} \right], \quad \text{for} \quad m > 0,$$

and

$$\frac{\sigma^2}{n + 1} + C \frac{\log(n + 1)}{\rho^3 [n + 1]^{1/2}} \left[\log(n + 1) + \frac{\rho}{\sigma} \right], \quad \text{for} \quad m = 0,$$

the number of roots of $F_n(z, \omega)$ in the circle Γ_m is at most $\log(\alpha n^3)/\log 2$. Considering circles Γ_i, $i = 0, 1, \ldots, k$, $p \log n$, we have the result that the number of roots inside all the circles is at most

$$[(k + 2)\log(\alpha n^3)]/\log 2 \le [(p \log n + 2) \log(\alpha n^3)]/\log 2$$
$$< \alpha(\log n)^2. \qquad (3.27)$$

The measure of the exceptional set is at most

$$\frac{\sigma^2(k + 2)}{n + 1} + C \frac{\log(n + 1)}{\rho^3[n + 1]^{1/2}} \left[\log(n + 1) + \frac{\rho}{\sigma}\right]$$

$$+ \frac{C\sqrt{2}}{\rho^3} \sum_{m=1}^{p \log n} \frac{\log^2 M(m)}{[M(m)]^{1/2}} + \frac{C\sqrt{2}}{\rho^2} \sum_{m=1}^{p \log n} \frac{\log M(m)}{M(m)}. \qquad (3.28)$$

To show that the measure of the exceptional set tends to zero as n tends to infinity, we first observe that the first two terms tend to zero as n tends to infinity. From the definition of $M(m)$, that is (3.24), we see that $M(m) > D2^m \log n$, where D is an absolute constant. Hence the last two terms in (3.28) are $O((\log \log n)^2/[\log n]^{1/2})$, which tends to zero as n tends to infinity.

To complete the proof of the theorem we now consider the interval $(0, \frac{1}{2})$. From (3.20) the number of roots of any $F_n(z, \omega)$ in the circle $|z| \le \frac{1}{2}$ is at most

$$\log \frac{\max_{|z| \le 1} |F_n(z, \omega)|}{|F_n(0, \omega)|} (\log 2)^{-1}.$$

Since $F_n(0, \omega) = a_0(\omega)$, the above expression is not defined if $\mathscr{P}\{a_0(\omega) = 0\} > 0$, for in this case $\mathscr{P}\{F_n(0, \omega) = 0\} > 0$. To circumvent this difficulty we take a circle with center r'' $(0 < r'' < \frac{1}{2})$ and radius $\frac{1}{2}$. The circle $|z - r''| \le 1 - r''$ is interior to the circle $|z| \le 1$. Hence, putting $z_0 = r''$, $r = \frac{1}{2}$ and $r' = 1 - r''$ in (3.20), we obtain the result that the number of roots of any $F_n(z, \omega)$ in the circle $\Gamma^* = \{z: |z - r''| \le \frac{1}{2}\}$ is at most

$$\log \frac{\max_{|z| \le 1} |F_n(z, \omega)|}{|F_n(r'', \omega)|} (\log 2(1 - r''))^{-1}.$$

Applying (3.22) we have

$$\max_{|z| \le 1} |F_n(z, \omega)| < \alpha n^2$$

except for a set of measure $\sigma^2/(n + 1)$. If we now apply Lemma 3.2 to $F_n(r'', \omega)$ by taking the subsequence s to be the sequence $1, 2, \ldots, \lambda_n$, where $\lambda_n = -\log n/\log r''$, we have

$$\mathcal{P}\left\{|F_n(r'', \omega)| \le \frac{1}{n}\right\} \le C \frac{\log \lambda_n}{\rho^3 \sqrt{\lambda_n}}\left[\log n + \frac{\rho}{\sigma n r'' \lambda_n}\right]$$

$$= C \frac{\log \lambda_n}{\rho^3 \sqrt{\lambda_n}}\left[\log n + \frac{\rho}{\sigma}\right]$$

$$= \varepsilon_{n'}$$

since $nr''\lambda_n = 1$. Since $\lambda_n \to \infty$, as $n \to \infty$, we have $\varepsilon_n \to 0$ as $n \to \infty$. Hence we can conclude that the number of roots inside Γ^* is at most

$$\log(\alpha n^3)/\log 2(1 - r'') < \alpha(\log n)^2, \tag{3.29}$$

except for a set of measure at most

$$\sigma^2/(n + 1) + \varepsilon_n, \tag{3.30}$$

but this tends to zero as $n \to \infty$. This completes the proof of the theorem.

We now prove the following theorem, which gives a lower bound for $N_n(R, \omega)$.

Theorem 3.9. *Let $F_n(x, \omega)$ be a random polynomial whose coefficients satisfy the conditions stipulated earlier, and let $\{\delta_n\}$ be a sequence of numbers tending to zero but such that $\delta_n \log n$ tends to infinity. Then, for $n \ge n_0$*

$$N_n(R, \omega) \ge \delta_n \log n.$$

The measure of the exceptional set tends to zero as n tends to infinity.[4]

The proof of Theorem 3.9 requires the use of three lemmas, the first of which is due to Berry [9] and Esseen [21], while the other two are given in [56].

Lemma 3.3. *Let $x_1, x_2, \ldots, x_n$ be a sequence of independent real-valued random variables with $\mathscr{E}\{x_i\} = 0$, variance β_i^2, and third absolute moment γ_i^3. Let $G_n(t)$ be the distribution function of $(1/\mu_n) \sum_{i=1}^n x_i$, where $\mu_n^2 = \sum_{i=1}^n \beta_i^2$, and let*

$$\Phi(t) = \frac{1}{\sqrt{2\pi}} \int_{-\infty}^t \exp[-\tfrac{1}{2}u^2] \, du.$$

---

[4] Theorem 3.9 can also be formulated as follows: *For each sequence of numbers $\{\delta_n\}$ such that $\delta_n \to 0$ but with $\delta_n \log n \to \infty$, $\mathcal{P}\{N_n(R, \omega) < \delta_n \log n\} \to 0$ as $n \to \infty$.*

Then

$$\sup_t |G_n(t) - \Phi(t)| \le 2\frac{\Lambda_n}{\mu_n},$$

where $\Lambda_n = \max_{1 \le i \le n} \lambda_i$, *and* $\lambda_i = \gamma_i^2/\beta_i^2$ *if* $\beta_i \ne 0$ *and zero otherwise.*

Before stating the next two lemmas we introduce some notation that will be used throughout the proof of Theorem 3.9. Let A and B be constants, with $A > 1$ and $0 < B < 1$, and let $\lambda = \{\lambda_n\}$ be a sequence of numbers tending to infinity with n. Let M be the integer defined by

$$M = [(8A\lambda e/B)^2] + 1, \tag{3.31}$$

and let k be the integer determined by the inequality

$$M^{2k} \le n < M^{2k+2}. \tag{3.32}$$

We will consider a random polynomial $F_n(x, \omega)$ at the points $x_m = 1 - (1/M^{2m})$, where $m = [\frac{1}{2}k] + 1, [\frac{1}{2}k] + 2, \ldots, k$. There will be $\frac{1}{2}k$ of these points if k is even, and $\frac{1}{2}(k + 1)$ points if k is odd. The random polynomial at the points x_m can be written as

$$F_n(x_m, \omega) = \sum_{k=0}^n a_k(\omega)x_m{}^k = U_m + R_{m'} \tag{3.33}$$

where

$$U_m = \sum_{k=M^{2m-1}+1}^{M^{2m+1}} a_k(\omega)x_m{}^k, \tag{3.34}$$

and

$$R_m = \sum_{k=0}^{M^{2m-1}} a_k(\omega)x_m{}^k + \sum_{k=M^{2m+1}+1}^n a_k(\omega)x_m{}^k. \tag{3.35}$$

We remark that U_m and U_{m+1} are independent random variables. Finally, we define

$$V_m = \frac{1}{2}\left(\sum_{k=M^{2m-1}+1}^{M^{2m+1}} x_m^{2k}\right)^{1/2}. \tag{3.36}$$

Lemma 3.4.

$$\tfrac{1}{2}V_m > \left| \sum_{k=M^{2m+1}+1}^n a_k(\omega)x_m{}^k \right|,$$

except for a set of measure at most

$$\frac{16A^2\sigma^2 e^2}{B^2} e^{-2M}.$$

Lemma 3.5.

$$\lambda\left(\sum_{k=0}^{M^{2m-1}} x_m^{2k}\right)^{1/2} > \left|\sum_{k=0}^{M^{2m-1}} a_k(\omega)x_m^{\ k}\right|$$

except for a set of measure at most

$$\left(\frac{2}{\pi}\right)^{1/2}\frac{\sigma}{\lambda}\exp\left\{-\frac{1}{2}\frac{\lambda^2}{\sigma^2}\right\} + \frac{4\tau^3 Ae}{3}\left(\frac{1}{M^{1/2}(k-1)}\right),$$

where σ^2 and τ^3 are the variance and third absolute moment of $a_k(\omega)$, respectively.

Proof of Theorem 3.9. We begin by estimating the following probability:

$$\mathcal{P}^* = \mathcal{P}\{(U_{2m} > V_{2m}, U_{2m+1} < -V_{2m+1})$$
$$\cup (U_{2m} < -V_{2m}, U_{2m+1} > V_{2m+1})\}$$
$$= \mathcal{P}\{U_{2m} > V_{2m}\}\mathcal{P}\{U_{2m+1} < -V_{2m+1}\}$$
$$+ \mathcal{P}\{U_{2m} < -V_{2m}\}\mathcal{P}\{U_{2m+1} > V_{2m+1}\}.$$

Let $\sigma_{2m}^2 = \mathcal{D}^2\{U_{2m}\}$ (the variance of U_{2m}); then from (3.35) and (3.36) we have

$$\sigma_{2m} = 2\sigma V_{2m}, \tag{3.37}$$

where $\sigma^2 = \mathcal{D}^2\{a_k(\omega)\}$. Now, let $G_{2m}(t)$ be the distribution function of U_{2m}/σ_{2m}. Then

$$\mathcal{P}\{U_{2m} < -V_{2m}\} = G_{2m}\left(-\frac{1}{2\sigma}\right)$$

$$\geq \Phi\left(-\frac{1}{2\sigma}\right) - \left|G_{2m}\left(-\frac{1}{2\sigma}\right) - \Phi\left(-\frac{1}{2\sigma}\right)\right|$$

Applying Lemma 3.3, and using (3.37) we have

$$\sup_t |G_{2m}(t) - \Phi(t)| \leq 2\tau^3/\sigma^2\sigma_{2m} \leq \tau^3/\sigma^3 V_{2m}.$$

Therefore

$$\mathcal{P}\{U_{2m} < -V_{2m}\} \geq \Phi(-1/2\sigma) - \tau^3/\sigma^3 V_{2m}.$$

Proceeding in the same way as above, estimates of the other three probabilities can be obtained. Combining these estimates we have

$$\mathscr{P}^* \geq [1 - \Phi(1/2\sigma) - \tau^3/\sigma^3 V_{2m}] \cdot [\Phi(-1/2\sigma) - \tau^3/\sigma^3 V_{2m+1}]$$
$$+ [\Phi(-1/2\sigma) - \tau^3/\sigma^3 V_{2m}] \cdot [1 - \Phi(-1/2\sigma) - \tau^3/\sigma^3 V_{2M+1}]. \tag{3.38}$$

From (3.36) we have

$$V_m \geq \frac{1}{2} \left(\sum_{k=M^{2m-1}+1}^{M^{2m}} x_m^{2k} \right)^{1/2}$$

$$\geq \frac{1}{2} M^m \left(1 - \frac{1}{M}\right)^{1/2} \left(1 - \frac{1}{M^{2m}}\right)$$

$$> \frac{1}{2} M^m B/eA. \tag{3.39}$$

Using the definition of the points x_m, we have $V_{2m} > \alpha M^k$; where α is an absolute constant. From the above estimate and (3.38) it follows that the probability $\mathscr{P}^*$ is greater than a quantity which tends to

$$2\Phi(-1/2\sigma)[1 - \Phi(1/2\sigma)]$$

as $n \to \infty$. This limit being positive, we can conclude that $\mathscr{P}^* > v$, where $v > 0$ is an absolute constant. Let sets E_m and F_m be defined as follows:

$$E_m = \{U_{2m} > V_{2m}, U_{2m+1} < -V_{2m+1}\}$$
$$F_m = \{U_{2m} < -V_{2m}, U_{2m+1} > V_{2m+1}\}.$$

From the estimation of $\mathscr{P}^*$ it follows that the measure of the set $E_m \cup F_m$ is greater than v.

Let η_m be a random variable such that it takes the value 1 on $E_m \cup F_m$ and zero elsewhere, i.e.,

$$\mathscr{P}\{\eta_m = 1\} = v_m > v,$$
$$\mathscr{P}\{\eta_m = 0\} = 1 - v_m. \tag{3.40}$$

The η_m are independent random variables with $\mathscr{E}\{\eta_m\} = v_m$ and $\mathscr{D}^2\{\eta_m\} = v_m - v_m^2$. We consider the sequence $\{\eta_m\}$ for $m = [\frac{1}{2}k] + 1, [\frac{1}{2}k] + 2, \ldots, k$. Now, let q be the total number of pairs (U_{2p}, U_{2p+1}) for which $[\frac{1}{2}k] + 1 \leq 2p < 2p + 1 \leq k$. There are $k - [\frac{1}{2}k]$ values of m; hence q must be at least euqal to $\frac{1}{2}(k - [\frac{1}{2}k] - 2)$.

Put $\eta = \sum_m \eta_m$, where the summation is to be taken over all q pairs. For $0 < \varepsilon < v$, an application of Chebychev's inequality gives

$$\mathscr{P}\{|\eta - \mathscr{E}\{\eta\}| \geq q\varepsilon\} \leq \mathscr{D}^2\{\eta\}/q^2\varepsilon^2 \leq \sum_m v_m/q^2\varepsilon^2 \leq 1/q\varepsilon^2.$$

But

$$q \geq \tfrac{1}{2}(k - [\tfrac{1}{2}k] - 2) \geq (k - \tfrac{1}{2}k - 2) = \tfrac{1}{4}(k - 4) \geq \alpha k.$$

Therefore, outside a set of measure at most α/k,

$$|\eta - \mathscr{E}\{\eta\}| < q\varepsilon;$$

that is

$$\eta - \mathscr{E}\{\eta\} > -q\varepsilon,$$

or

$$\eta > \mathscr{E}\{\eta\} - q\varepsilon = \sum_m v_m - q\varepsilon > q(\lambda - \varepsilon) \geq \alpha k.$$

Thus far we have shown that outside a set of measure at most α/k the following events occur: either $U_{2m} > V_{2m}$ and $U_{2m+1} < -V_{2m+1}$, or $U_{2m} < -V_{2m}$ and $U_{2m+1} > V_{2m+1}$ for at least αk values of m.

Let us now consider the random variable R_m. Applying Lemmas 3.4 and 3.5, we have that for any given m

$$|R_m| < \tfrac{1}{2}V_m + \lambda \left(\sum_{k=0}^{M^{2m-1}} x_m^{2k} \right)^{1/2} < \tfrac{1}{2}V_m + \lambda M^{m-1/2} < V_m,$$

except for a set Γ_m of measure at most

$$\tfrac{1}{2} \exp\{-C_1\lambda^2\} + (C_2/2\lambda^k),$$

where C_1 and C_2 are constants which can be determined in terms of the constants A, B, and σ.

Put

$$\phi_n = \begin{cases} 0, & \text{if } |R_m| < V_m \text{ and } |R_{2m+1}| < V_{m+1}. \\ 1, & \text{otherwise.} \end{cases} \tag{3.41}$$

Then if $\eta_m - \eta_m \phi_m = 1$ there is a root of the random polynomial in the interval (x_{2m}, x_{2m+1}). Hence the number of roots in (x_{2m_0}, x_{2k+1}), where $m_0 := [\tfrac{1}{2}k] + 1$, must exceed

$$\sum_{m=m_0}^{k} (\eta_m - \eta_m \phi_m).$$

But

$$\mathcal{E}\{\sum_m \eta_m \phi_m\} = \sum_m \mathcal{E}\{\eta_m \phi_m\} \leq \sum_m \mathcal{E}\{\phi_m\}$$

$$\leq (k+1)[\exp\{-C_1\lambda^2\} + C_2\lambda^{-k}].$$

Therefore,

$$\sum_{m=m_0}^{k} \eta_m \phi_m \leq k[\exp\{-\tfrac{1}{2}C_1\lambda^2\} + C_2\lambda^{-k+1}]$$

except for a set of measure at most $1/\lambda$. Hence, from the final statement in the proof of the relations between the U's and the V's, we have

$$\sum_{m=m_0}^{k} (\eta_m - \eta_m \phi_m) \geq k[\alpha - \exp\{-\tfrac{1}{2}C_1\lambda^2\} - C_2\lambda^{-k+1}]$$

outside a set of measure at most $\alpha/k + 1/\lambda$. The sequence $\lambda = \{\lambda_n\}$ and k are connected by (3.31) and (3.32), and we can find a constant C_3 such that if $\lambda = \{\lambda_n\} = \exp[C_3/\delta_n]$, then the measure of the exceptional set tends to zero as $n \to \infty$. Furthermore, because of the relationship between k and λ we have $k > C\delta_n \log n$ where C is an appropriate constant. If $\delta_n \log n \to \infty$ as $n \to \infty$, then the statement of the theorem follows.

Combining the results of Theorems 3.8 and 3.9, we have

$$\delta_n \log n \leq N_n(R, \omega) \leq \alpha(\log n)^2; \tag{3.42}$$

or, more precisely,

$$\lim_{n \to \infty} \mathcal{P}\{\delta_n < N_n(R, \omega)/\log n < \alpha \log n\} = 1 \tag{3.43}$$

for a certain constant $\alpha > 0$ and each sequence of numbers $\{\delta_n\}$ such that $\delta_n \to 0$ but with $\delta_n \log n \to \infty$. This result is a strict generalization of the results of Littlewood and Offord (Theorems 3.1 and 3.2).

D. ROOTS OF RANDOM ALGEBRAIC POLYNOMIALS WITH COMPLEX COEFFICIENTS

All of the studies referred to thus far have been concerned with the number, or average number, or both of real roots of algebraic polynomials $F_n(x, \omega)$, where $x \in R$, and, with one exception, the coefficients $a_k(\omega)$ are real-valued random variables. The one exception is Theorem 3.3 due to Littlewood and Offord, which assumed the coefficients $a_k(\omega)$ to be complex-valued random variables. In this section we summarize some results due to Dunnage and Arnold on the number and expected number of real or complex roots of algebraic polynomials whose coefficients are complex-valued random variables.

We now consider an algebraic polynomial of the form

$$F_n(z, \omega) = \sum_{k=0}^{n} a_k(\omega)z^k, \qquad \text{where} \qquad a_k(\omega) = \alpha_k(\omega) + i\beta_k(\omega),$$

($\alpha_k(\omega)$ and $\beta_k(\omega)$ are centered real-valued random variables) and $z = x + iy$. The first result we consider is due to Dunnage [16], and is a general form of Theorem 3.3. In the statement of the theorem we will use the following notation: $s_k^2 = \mathscr{E}\{\alpha_k^2(\omega)\}$, $u_k^2 = \mathscr{E}\{\beta_k^2(\omega)\}$, $S_k^3 = \mathscr{E}\{|\alpha_k(\omega)|^3\}$, $U_k^3 = \mathscr{E}\{|\beta_k(\omega)|^3\}$, and $v_k^2 = \mathscr{E}\{|a_k(\omega)|^2\} = s_k^2 + u_k^2$. Also, we denote by $G_k(\xi)$ the distribution function of $|a_k(\omega)|$.

Theorem 3.10. *Let $N_n(R, \omega)$ denote the number of real roots of $F_n(x, \omega)$* $= \sum_{k=0}^{n} a_k(\omega)x^k$. *Then, for n sufficiently large,*

$$\mathscr{P}\left\{N_n(R, \omega) \geq 10 \log n \left[\log \frac{M}{v_0\, v_n} + 2(\log n)^5\right]\right\}$$

$$\leq \frac{C}{\gamma_n^9} \frac{\log \log n}{\log n} + 2G_0\left(\frac{v_0}{n}\right) + 2G_n\left(\frac{v_n}{n}\right),$$

where $M = \sum_{k=0}^{n} v_k$, C is an absolute constant, and

$$\gamma_n = \min(v_k/\max(S_k, U_k)).$$

The minimum being taken over all values of k for which $v_k \neq 0$.

We remark that the estimate of $N_n(R, \omega)$ given by the above theorem includes not only roots at zero, but also conventional roots at infinity.

We also remark that Theorem 3.2 of Littlewood and Offord does not have an analog at the level of generality of Theorem 3.10. To illustrate this point, suppose $F_n(x, \omega) = \sum_{k=0}^{n} a_k(\omega)x^k$ has a real root. This implies that the polynomials $\sum_{k=0}^{n} \alpha_k(\omega)x^k$ and $\sum_{k=0}^{n} \beta_k(\omega)x^k$ have a common root; and the elimination of x leads to the equation

$$\Phi(\alpha_0, \alpha_1, \ldots, \alpha_n, \beta_0, \beta_1, \ldots, \beta_n) = 0, \qquad (3.44)$$

where Φ is a polynomial in the $2n + 2$ variables. Now suppose that the $2n + 2$ random variables are independent, and that the probability distribution of each can be defined by a density function; then their joint probability distribution admits a $(2n + 2)$-dimensional density function, $h(x)$ say, where $x \in E_{2n+2}$. The probability that (3.44) holds is given by

$$\int_{\Gamma} h(x)\, dx,$$

where Γ is the curve derived from (3.44) by considering the α_k and β_k to be Cartesian-coordinate variables. The above integral is clearly zero, since the Lebesgue measure of Γ is zero. Therefore, the probability that $F_n(x, \omega)$ has any real roots is, in this case, zero.

Another interesting result due to Dunnage concerns the multiplicity of the roots of a random algebraic polynomial

$$F_n(z, \omega) = \sum_{k=0}^{n} \alpha_k(\omega) z^k, \qquad z \in Z.$$

Theorem 3.11. *Consider the random polynomial $F_n(z, \omega)$, $z \in Z$, where the coefficients $\alpha_k(\omega)$ are independent real-valued random variables whose probability distributions admit density functions. Then, the probability that $F_n(z, \omega)$ has any multiple roots anywhere in the complex plane is zero.*

Proof. The proof is based on the same sort of argument used above. Suppose $F_n(z, \omega)$ has a multiple root at $z = \xi_0$. Then $F_n(\xi_0, \omega) = F_n'(\xi_0, \omega) = 0$; and the elimination of ξ_0 leads to the condition $\Phi(\alpha_0, \alpha_1, \ldots, \alpha_n) = 0$, where Φ is a polynomial in the coefficients $\alpha_k(\omega)$. The remainder of the proof follows the argument used previously.

In closing this section, we state some results due to Arnold [3, 4] concerning $v_n(B) = \mathscr{E}\{N_n(B, \omega)\}$, $B \in \mathscr{B}(Z)$, for random polynomials with complex coefficients.

Theorem 3.12. *If the real parts of the coefficients $a_k(\omega)$ (i.e., $\alpha_k(\omega)$ and $\beta_k(\omega)$) are independent and normally distributed $(0, \sigma)$ real-valued random variables, then*

(i) $v_n(B)$ *is m_2-continuous on $\mathscr{B}(Z)$, where m_2 denotes Lebesgue measure on $\mathscr{B}(Z)$, and has the density ($z = re^{i\theta}$)*

$$\rho_n(r, \theta) = \begin{cases} \dfrac{1}{\pi} \dfrac{1 - \Phi_n^{2}(r)}{(1 - r^2)^2}, & \text{for} \quad r \neq 1 \\[2ex] \dfrac{1}{\pi} \dfrac{n(n+2)}{12}, & \text{for} \quad r = 1. \end{cases} \tag{3.45}$$

In (3.45)

$$\Phi_n(r) = (n + 1)r^n\left(\frac{1 - r^2}{1 - r^{2n+2}}\right).$$

(ii) $v_n(|z| < r) = \begin{cases} \dfrac{r^2}{1 - r^2} - (n + 1)\dfrac{r^{2n+2}}{1 - r^{2n+2}}, & r \in (0, 1) \\[4mm] \dfrac{n}{2}, & r = 1. \end{cases}$ (3.46)

(iii)   *for all r > 0,*

$$v_n(|z| > r) = v_n(|z| < 1/r).$$ (3.47)

Theorem 3.12 is a generalization of the result due to Rice (Section III.A), and (3.45) is the complex analog of Kac's result. We remark that the above results yield the obvious relation $v_n(|z| < \infty) = n$.

Arnold has also shown that under the conditions of Theorem 3.12 the following asymptotic results obtain:

$$\lim_{n \to \infty} v_n(|z| < r) = \frac{r^2}{1 - r^2}, \qquad r \in (0, 1),$$ (3.48)

$$v_n(|z| < 1 - n^{-p}) \sim n^p/2, \qquad p \in (0, 1),$$ (3.49)

$$v_n(|z| < 1 - cn^{-1}) \sim \frac{n}{2}\left(\frac{1}{c} - \frac{2}{e^{2c} - 1}\right), \qquad c > 0,$$ (3.50)

$$v_n(|z| < 1 - \varphi_n n^{-1}) \sim n/2, \qquad \varphi_n \geq 0, \qquad \varphi_n \to 0.$$ (3.51)

IV. Distribution of the Roots of a Random Algebraic Polynomial

A. INTRODUCTION

In this section we consider some studies devoted to the following problem: Given the joint probability distribution (or density) of the coefficients $a_0(\omega), a_1(\omega), \ldots, a_n(\omega)$ of a random polynomial

$$F_n(z, \omega) = a_0(\omega) + a_1(\omega)z + \cdots + a_n z^n,$$ (4.1)

determine the joint probability distribution (or density) of the roots (or solutions) $\xi_1(\omega), \xi_2(\omega), \ldots, \xi_n(\omega)$ of the random algebraic equation

$F_n(z, \omega) = 0$. In the general case, $a_k(\omega) = \alpha_k(\omega) + i\beta_k(\omega)$ and $\xi_k(\omega) = \gamma_k(\omega) + i\lambda_k(\omega)$, where $\alpha_k(\omega)$, $\beta_k(\omega)$, $\gamma_k(\omega)$, and $\lambda_k(\omega)$ are real-valued random variables. Hence, the problem of determining the distribution function (or density) of the roots given the distribution function (or density) of the coefficients involves the development of a transformation theory of probability measures on the $(2n + 2)$-dimensional space of coefficients into probability measures on the $(2n + 2)$-dimensional space of solutions. We remark that it is necessary to consider the cases of real and complex coefficients separately. Also, when the coefficients are real-valued random variables the solutions are, in general, complex-valued random variables, and therefore require separate treatment.

In Section IV.B we consider the problem of determining the distribution function of the solutions of random linear and quadratic equations and present some explicit results in these cases. For a detailed treatment of some random algebraic equations of degree 3 and 5, we refer to Gaede [26, 26a]. Section IV.C is devoted to a result of Girshick [28], and Hammersley [31], on the distribution of the roots of a random polynomial with complex coefficients. Finally, in Section IV.D we outline Hammersley's approach to the problem of determining the roots of a random algebraic equation based on the so-called condensed distribution of the roots.

B. DISTRIBUTION OF THE ROOTS OF RANDOM LINEAR AND QUADRATIC EQUATIONS

1. *Random linear equations.*

In this section we consider the problem of determining the distribution (or density) of the solution of the random linear equation

$$F_1(x, \omega) = a_0(\omega) - a_1(\omega)x = 0. \qquad (4.2)$$

One of the early results applicable to the above equation is due to Geary [27], who studied the distribution of the ratio of two normally distributed real-valued random variables. Let $y_0(\omega)$ and $y_1(\omega)$ be normally distributed real-valued random variables with means 0 and standard deviations σ_0 and σ_1, respectively. Let ρ denote the correlation coefficient between y_1 and y_2. If we put

$$r = \frac{y_0 + \tau_0}{y_1 + \tau_1},$$

then Geary's result is that

$$\phi = \frac{\tau_0 r - \tau_1}{(\sigma_0 r^2 - 2\rho\sigma_0\sigma_1 r + \sigma_1{}^2)^{1/2}},$$

where τ_0 and τ_1 are constants, is normally distributed with mean 0 and standard deviation 1. The above result is valid for $\tau_1 \geq 3\sigma_0$, for under this condition $y_1 + \tau_1$ does not assume negative values. In the field of bioassay this result is known as *Fieller's theorem* (cf. [24, pp. 27–29]). It is clear that Geary's result can be used to obtain the distribution of the single root of Eq. (4.2) when $a_0(\omega)$ and $a_1(\omega)$ are normally distributed real-valued random variables with means 0 and standard deviations σ_0 and σ_1, respectively, and with $a_1(\omega) \neq 0$ almost surely; for in this case $\tau_0 = \tau_1 = 0$, and the probability density of the solution $\xi(\omega) = a_0(\omega)/a_1(\omega)$ is

$$f(\xi) = \frac{\sigma_0\sigma_1(1 - \rho^2)^{1/2}}{\pi}(\sigma_0\xi^2 - 2\rho\sigma_0\sigma_1\xi + \sigma_1{}^2)^{-1}. \qquad (4.3)$$

Following Kabe [33], we now utilize multivariate normal distribution theory to show that the distribution of the root $\xi(\omega) = a_0(\omega)/a_1(\omega)$ of the random linear equation (4.2) can be expressed in terms of known functions. Let the coefficients $a_0(\omega)$ and $a_1(\omega)$ have a joint bivariate normal distribution with density

$$g(\mathbf{a}) = (|B|^{1/2}/2\pi)\exp\{-\tfrac{1}{2}(\mathbf{a} - \boldsymbol{\mu})'B(\mathbf{a} - \boldsymbol{\mu})\},$$

where $\mathbf{a}' = (a_0, a_1)$, $\boldsymbol{\mu}' = (\mu_0, \mu_1)$, with $\mu_i = \mathscr{E}\{a_i\}$, and

$$B = \begin{pmatrix} b_{00} & b_{01} \\ b_{10} & b_{11} \end{pmatrix}$$

is a positive definite (symmetric) matrix. As is well-known $B^{-1} = \Sigma$, where Σ is the covariance matrix

$$\Sigma = \begin{pmatrix} \sigma_0{}^2 & \sigma_0\sigma_1\rho \\ \sigma_1\sigma_0\rho & \sigma_1{}^2 \end{pmatrix}$$

and ρ is the correlation coefficient between a_0 and a_1. A routine calculation yields

$$\frac{[\boldsymbol{\mu}'B\mathbf{a}]^2}{\mathbf{a}'B\mathbf{a}} = \frac{[(\mu_0 b_{00} + \mu_1 b_{01})\xi + (\mu_0 b_{01} + \mu_1 b_{11})]^2}{(b_{00}\xi^2 + 2b_{01}\xi + b_{11})};$$

or

$$\frac{\mu' B a}{[a' B a]^{1/2}} = \frac{(\mu_0 b_{00} + \mu_1 b_{01})\xi + (\mu_0 b_{01} + \mu_1 b_{11})}{[b_{00}\xi^2 + 2b_{01}\xi + b_{11}]^{1/2}} = t,$$

say.

It is known that (cf. [53, formula (3.5-9)])

$$f(\xi' B\xi, \psi\xi)\, d\xi$$

$$= \tfrac{1}{2} C(N-1)|B|^{-1/2}(\psi' B^{-1}\psi)^{-1/2} f(u, v) \left[u - \frac{v^2}{\psi' B\psi}\right]^{(N-3)/2}, \quad (4.5)$$

where the integral is taken over the range of $\xi' B\xi = u$ and $\psi\xi = v$. In (4.5) ξ and ψ are N component column vectors and $C(N)$ denotes the surface area of the unit sphere in N dimensions. We assume that the function f is such that the integral (4.5) exists. Obviously, if the function is a suitable density function, then the right-hand side of (4.5) represents the joint density function of u and v.

The density function $g(\mathbf{a})$ as given by (4.4) can be rewritten as

$$g(\mathbf{a}) = (|B|^{1/2}/2\pi) \exp\{-\tfrac{1}{2} a' B a + \mu' B a - \tfrac{1}{2}\mu' B\mu\}. \quad (4.6)$$

If we now put $u = a' B a$ and $v = \mu' B a$, and use (4.5), it follows from (4.6) that the joint density function of u and v is given by

$$g(u, v) = [(\mu' B\mu)^{-1/2}/2\pi] \exp\{-\tfrac{1}{2} u + v - \mu' B\mu\}[u - (v^2/\mu' B\mu)]^{-1/2}$$

$$= [(\mu' B\mu)^{-1/2}/2\pi] \exp\{-\tfrac{1}{2}\mu' B\mu\} \exp\{-\tfrac{1}{2} u\}$$

$$\cdot [u - (v^2/\mu' B\mu)]^{-1/2} \cdot \sum_{k=0}^{\infty} v^{2k}/(2k)!. \quad (4.7)$$

The joint density of u and t can be obtained from (4.7) by setting $v = u^{1/2} t$; and integration of the resulting expression with respect to u yields the density function of t:

$$h(t) = \frac{(\mu' B\mu)^{-1/2}}{\pi} \exp\{-\tfrac{1}{2}\mu' B\mu\}$$

$$\cdot \left[1 - \frac{t^2}{\mu' B\mu}\right]^{-1/2} \cdot \sum_{k=0}^{\infty} \frac{\Gamma(k+1)\Gamma(\tfrac{1}{2}) t^{2k}}{k!\,\Gamma(k+\tfrac{1}{2}) 2^k}$$

$$= (\mu' B\mu)^{-1/2} \exp\{-\tfrac{1}{2}\mu' B\mu\}$$

$$\cdot \left[1 - \frac{t^2}{\mu' B\mu}\right]^{-1/2} {}_1F_1(1, \tfrac{1}{2}; t^2/2). \quad (4.8)$$

In (4.8) $_1F_1(1, \frac{1}{2}; t^2/2)$ denotes the hypergeometric function

$$_1F_1(\alpha, \beta; x) = 1 + \frac{\alpha x}{\beta} + \frac{\alpha(\alpha + 1)x^2}{\beta(\beta + 1)2!} + \cdots.$$

From $h(t)$, the density function of $\xi(\omega)$, the solution of Eq. (4.2) can be found without difficulty; hence the distribution function of $\xi(\omega)$ can be obtained in terms of known functions.

2. *Random quadratic equations.*

Consider the random quadratic equation

$$F_2(x, \omega) = a_0{}^2(\omega) + 2a_1(\omega)x + x^2. \qquad (4.10)$$

We will assume that the coefficients $a_0(\omega)$ and $a_1(\omega)$ in Eq. (4.10) are real-valued random variables with joint bivariate normal density function

$$g(a_0, a_1) = [(b_{00} b_{11})^{1/2}/2\pi]\exp\{-\tfrac{1}{2}b_{00} a_1{}^2 - \tfrac{1}{2}b_{11}(a_0 - \mu_0)^2; \qquad (4.11)$$

hence we are assuming that $a_0(\omega)$ and $a_1(\omega)$ are independent, with $\mu_1 = 0$, and $\sigma_0 = \sigma_1 = 1$. The roots of Eq. (4.10) are

$$\begin{aligned} \xi_1(\omega) &= -a_1(\omega) + [a_1{}^2(\omega) - a_0{}^2(\omega)]^{1/2}, \\ \xi_2(\omega) &= -a_1(\omega) - [a_1{}^2(\omega) - a_0{}^2(\omega)]^{1/2}. \end{aligned} \qquad (4.12)$$

We observe that $\mathcal{P}\{\xi_1(\omega) = \xi_2(\omega)\} = 0$ and $\mathcal{P}\{\xi_1(\omega)$ and $\xi_2(\omega)$ real$\} = \mathcal{P}\{\xi_1(\omega)$ and $\xi_2(\omega)$ complex$\} = \frac{1}{2}$. We will restrict out attention to the *distribution of real roots* of Eq. (4.10); hence we assume that $a_1{}^2(\omega) - a_0{}^2(\omega) > 0$ almost surely.

Under the above assumptions Kabe has shown that the joint characteristic function of a_1 and $a_1{}^2 - a_0{}^2 = v$ is

$$\begin{aligned} \psi(t_1, t_2) &= \int_{-\infty}^{\infty} \int_{-\infty}^{\infty} \exp\{it_1 a_1 + it_2(a_1{}^2 - a_0{}^2)\} \\ &\quad \cdot g(a_0, a_1)\, da_0\, da_1 \\ &= \frac{(b_{00} b_{11})^{1/2}}{(b_{00} - 2it_2)(b_{11} + 2it_2)} \\ &\quad \cdot \exp\left\{-\frac{1}{2}b_{11}\mu_0{}^2 + \frac{1}{2}\frac{(it)^2}{(b_{00} - 2it_2)} \right. \\ &\quad \left. + \frac{(\mu_0 b_{11})^2}{(b_{00} - 2it_2)(b_{11} + 2it_2)}\right\}. \end{aligned} \qquad (4.13)$$

Inversion of (4.13) yields

$$g(a_1, v) = \frac{(b_{00} b_{11})^{1/2}}{(2\pi)^{1/2}} \exp\{-\tfrac{1}{2} b_{00} a_1^2 - \tfrac{1}{2} b_{11} \mu_0^2\}$$

$$\times \frac{1}{2\pi} \int_{-\infty}^{\infty} \frac{e^{-it_2}(v - a_1^2)}{(b_{00} - 2it_2)^{1/2}(b_{11} + 2it_2)}$$

$$\times \sum_{k=0}^{\infty} \frac{(\mu_0 b_{11})^{2k}}{(b_{00} - 2it_2)^k (b_{11} + 2it_2)^k (k!)} \, dt_2. \qquad (4.14)$$

The integral in (4.14) can be evaluated by using a known formula ([18, p. 119]); and we find that

$$g(a_1, v) = \frac{(b_{00} b_{11})^{1/2}}{(2\pi)^{1/2}} \exp\{-\tfrac{1}{2}(b_{00} + b_{11})a_1^2 + \tfrac{1}{4}(b_{11} - b_{00})v\}$$

$$\times \sum_{k=0}^{\infty} (a_1^2 - v)^{k-(3/4)} \frac{(b_{00} + b_{11})^{-(2k+(3/2))/2}}{2^{(2k+(3/2))/2}(k!)^2}$$

$$\times W_{1/4, \, -k-(1/4)}(-\tfrac{1}{2}(b_{00} + b_{11})(v - a_1^2)), \qquad (4.15)$$

where $W_{\alpha, \beta}(x)$ is the Whittaker confluent hypergeometric function (cf. [60, Chapter XVI]). From (4.15) it follows that the joint density of a_1 and $v^{1/2}$ is

$$g(a_1, v^{1/2}) = 2v^{1/2} g(a_1, v); \qquad (4.16)$$

hence we see that the joint density function of $\xi_1(\omega)$ and $\xi_2(\omega)$ is given by

$$g(\xi_1, \xi_2) = (\xi_1 - \xi_2) \frac{(b_{00} b_{11})^{1/2}}{(2\pi)^{1/2}} \exp\{-\tfrac{1}{2} \mu_0^2 b_{11}$$

$$- \tfrac{1}{16}(b_{00} + b_{11})(\xi_1 + \xi_2) + \tfrac{1}{16}(b_{11} - b_{00})(\xi_1 - \xi_2)^2\}$$

$$\times \sum_{k=0}^{\infty} \frac{(\xi_1 \xi_2)^{k-(3/4)}(b_{00} + b_{11})^{-(2k+(3/2))/2}}{2^{(2k+(3/2))/2}(k!)^2}$$

$$\times W_{1/4, \, -k-(1/4)}(\tfrac{1}{2}(b_{00} + b_{11})\xi_1 \xi_2). \qquad (4.17)$$

Finally, the distribution of the roots of the random quadratic equation (4.10) can be obtained from (4.17) in terms of series of Kummer's confluent hypergeometric functions.

Hamblen [30], in a detailed study of random quadratic equations, considered the equation

$$F_2(x, \omega) = a_0(\omega) - a_1(\omega)x + x^2 = 0, \qquad (4.18)$$

where $a_0(\omega)$ and $a_1(\omega)$ are real-valued random variables with joint density function $g(a_0, a_1)$. The joint density of two real roots was obtained, and in the case of complex roots the joint density of real and imaginary parts was obtained. Detailed results are given when $g(a_0, a_1)$ is (i) bivariate normal and (ii) gamma.

C. Distribution of the Roots of a Random Polynomial with Complex Coefficients

Consider a random algebraic polynomial of degree n of the form

$$F_n(z, \omega) = z^n - a_1(\omega)z^{n-1} + \cdots + (-1)^n a_n(\omega), \qquad (4.19)$$

where the coefficients $a_k(\omega)$, $k = 1, 2, \ldots, n$, are complex-valued random variables, the real and imaginary parts ($\alpha_k(\omega)$ and $\beta_k(\omega)$, respectively) of which are independent, normally distributed real-valued random variables with mean 0 and standard deviation σ. Let $\xi_1(\omega)$, $\xi_2(\omega)$, $\ldots$, $\xi_n(\omega)$ denote the roots of $F_n(z, \omega)$. Now, the relationship between the roots and coefficients of $F_n(z, \omega)$ is given by

$$a_1(\omega) = \sum_{k=1}^{n} \xi_k(\omega),$$

$$a_2(\omega) = \sum_{i<j} \xi_i(\omega)\xi_j(\omega), \qquad (4.20)$$

$$a_n(\omega) = \prod_{k=1}^{n} \xi_k(\omega).$$

Hence the coefficients $a_k(\omega)$ are, for every fixed $\omega \in \Omega$, analytic functions of the roots. In order to find the joint probability distribution of the real and imaginary parts of the roots (i.e., $\gamma_k(\omega)$ and $\lambda_k(\omega)$, respectively) it is necessary to obtain the real Jacobian D of the transformation defined by (4.20). We first state a lemma on complex Jacobians which will enable us to determine D.

Lemma 4.1. *Consider n analytic functions defined by*

$$u_k = v_k + iw_k = \Phi_k(z_1, z_2, \ldots, z_n), \qquad k = 1, 2, \ldots, n, \qquad (4.21)$$

where $z_k = x_k + iy_k$. Let d denote the complex Jacobian of the transformation defined by (4.21); that is

$$d = \left| \frac{\partial(u_1, u_2, \ldots, u_n)}{\partial(z_1, z_2, \ldots, z_n)} \right|. \qquad (4.22)$$

*Furthermore, let D denote the real Jacobian of the 2n real variables
defined by the equations*

$$v_k = v_k(x_1, y_1, x_2, y_2, \ldots, x_n, y_n),$$
$$w_k = w_k(x_1, y_1, x_2, y_2, \ldots, x_n, y_n); \tag{4.23}$$

that is

$$D = \left| \frac{\partial(v_1, w_1, v_2, w_2, \ldots, v_n, w_n)}{\partial(x_1, y_1, x_2, y_2, \ldots, x_n, y_n)} \right|. \tag{4.24}$$

Then $D = |d|^2$, i.e., D equals the square of the modulus of d.

For a proof of the above lemma we refer to Girshick [28] and
Hammersley [31].

Applying the lemma to the transformation defined by (4.20), we find

$$d = \begin{vmatrix} \dfrac{\partial a_1}{\partial z_1} & \dfrac{\partial a_1}{\partial z_2} & \cdots & \dfrac{\partial a_1}{\partial z_n} \\ \vdots & \vdots & \vdots & \vdots \\ \dfrac{\partial a_n}{\partial z_1} & \dfrac{\partial a_n}{\partial z_2} & \cdots & \dfrac{\partial a_n}{\partial z_n} \end{vmatrix}. \tag{4.25}$$

The value of d as given by (4.25) is

$$d = \sum_{i=1}^{n} \sum_{j=i+1}^{n} (z_i - z_j). \tag{4.26}$$

Therefore

$$D = |d|^2 = \sum_{i=1}^{n} \sum_{j=i+1}^{n} |z_i - z_j|^2. \tag{4.27}$$

If we now assume that the real and imaginary parts of the coefficients
$a_k(\omega)$ are independent, normally distributed real-valued random
variables with mean 0 and standard deviation σ, then their density is
given by

$$\left(\frac{1}{(2\pi)^{1/2} \sigma} \right)^{2n} \exp\left\{ -\frac{1}{2\sigma^2} \sum_{k=1}^{n} a_k \bar{a}_k \right\}, \tag{4.28}$$

where $\bar{a}_k$ denotes the conjugate of a_k. Utilizing (4.27), a straightforward calculation yields the joint density of the real and imaginary

parts of the roots $\xi_k(\omega)$ of (4.19)

$$\left(\frac{1}{(2\pi)^{1/2}\,\sigma}\right)^{2n}\exp\left\{-\frac{1}{2\sigma^2}\left[\sum_{k=1}^{n}z_k\sum_{k=1}^{n}\bar{z}_k+\cdots\right.\right.$$

$$\left.\left.+\,z_1\bar{z}_1\cdots z_n\bar{z}_n\right]\right\}\times\sum_{i=1}^{n}\sum_{j=i+1}^{n}|z_i-z_j|^2. \qquad (4.29)$$

For other studies which utilize the transformation (4.20) we refer to the papers of Gaede [26] and Hamblen [30].

D. CONDENSED DISTRIBUTION OF THE ROOTS OF A RANDOM POLYNOMIAL

In a fundamental paper on the roots of random polynomials, Hammersley [31] formulated an approach to the general problem of determining the distribution of the roots which did not utilize the notion of a joint distribution or density (as considered in Section IV.C), but was based on a new concept, namely the *condensed distribution of the roots*.

Consider Eq. (4.1) where the coefficients $a_k(\omega)$ are complex-valued random variables with an arbitrary distribution function

$$G(\mathbf{a})=G(\alpha_0,\ldots,\alpha_n,\beta_0,\ldots,\beta_n).$$

We will assume (i) that $G(a)$ possesses a continuous density function

$$g(\mathbf{a})=\frac{\partial^{2n+2}G(\mathbf{a})}{\partial\alpha_0\cdots\partial\alpha_n\,\partial\beta_0\cdots\partial\beta_n},$$

and (ii) that all moments of the coefficients $a_k(\omega)$ exist.

We now introduce some terminology. Let $x(\omega)$ be a measurable mapping of a probability space $(\Omega, \mathscr{A}, \mu)$ into a probability space $(\mathfrak{X}, \mathscr{B}, v)$, where $\mathfrak{X}$ is a finite-dimensional Euclidean space, $\mathscr{B}$ is the σ-algebra of Borel sets in $\mathfrak{X}$, and $v(B)=\mu(\{\omega: x(\omega)\in B, B\in\mathscr{B}\})$ with $v(\mathfrak{X})=1$. An *n-valued function* $y=y(x)$ mapping $\mathfrak{X}$ into a finite dimensional Euclidean space $\mathfrak{Y}$ is a set of n points (not necessarily distinct) in $\mathfrak{Y}$ corresponding to each given $x\in\mathfrak{X}$. An *indexing* of $y(x)$ is a system of n 1-valued functions $y_i(x)$, $i=1, 2, \ldots, n$, such that, for a given $x\in\mathfrak{X}$, the n points $y(x)$ coincide with the n points $y_i(x)$, with due respect for multiplicity. A function $y(x)$ is called an *n-valued Borel measurable function* if there exists at least one such indexing in which each $y_i(x)$ is a Borel measurable function; and the particular indexing is called a *Borel measurable indexing*. Finally, we have the following definition.

Definition 4.1. If $y_i(x)$, $i = 1, 2, \ldots, n$ is a Borel measurable indexing of an n-valued Borel measurable function, and if x is a 1-valued random variable specified by a measure $v(B)$, the 1-valued random variable $y^*(x)$ in $\mathfrak{Y}$ specified by the measure

$$v^*(C) = \frac{1}{n} \sum_{i=1}^{n} v[y_i^{-1}(C)], \qquad (4.30)$$

where C is a Borel subset of $\mathfrak{Y}$, is called the *condensation* of $y(x)$, and v^* the *condensed distribution* of $y^*(x)$.

Let E_{2n+2} denote the $(2n + 2)$-dimensional Euclidean space consisting of the points $\mathbf{a} = (\alpha_0, \alpha_1, \ldots, \alpha_n, \beta_0, \beta_1, \ldots, \beta_n)$ and let Z denote the complex plane. Let dz_k, $k = 1, 2, \ldots, m$, denote the rectangle $x_k < x \leq x_k + dx_k$, $y_k < y \leq y_k + dy_k$, where the $z_k = x_k + iy_k$ are prescribed; and let $P(z_1, z_2, \ldots, z_m)\, dz_1\, dz_2 \cdots dz_m$ denote the probability that the random algebraic equation $F_n(z, \omega) = 0$ has *exactly* one root in each of the rectangles dz_k. For fixed $z = x + iy$, the equations

$$\mathrm{Re}[F_n(z, \omega)] = \mathrm{Im}[F_n(z, \omega)] = 0, \qquad (4.31)$$

define the $2n$-dimensional subspace $\mathscr{F}_z$ in E_{2n+2}; and as z varies, the subspaces $\mathscr{F}_z$ develop a twisted regulus. The generator of the regulus lying in $\mathscr{F}_z$ is a $(2n - 2)$-dimensional subspace $\mathscr{H}_z$ with associated equations

$$\mathrm{Re}[F_n(z, \omega)] = \mathrm{Im}[F_n(z, \omega)]$$
$$= \mathrm{Re}[F_n'(z, \omega)] = \mathrm{Im}[F_n'(z, \omega)] = 0. \qquad (4.32)$$

In order to write down an expression for the condensed distribution of the roots of $F_n(z, \omega) = 0$, we argue as follows. If $F_n(z, \omega) = 0$ has *at least* one root in each of the rectangles dz_k, $k = 1, 2, \ldots, m$, then $a(\omega)$ must lie at the intersection of the m $2n$-dimensional subspaces $\mathscr{F}_{z_k}$. Furthermore, if one of the rectangles, say dz_j, contains more than one root, then $\mathbf{a}(\omega)$ must also lie on $\mathscr{H}_{z_j}$. By assumption the density function $g(\mathbf{a})$ is continuous; hence it follows that the probability that each of the rectangles dz_k contains exactly one root differs by terms of higher order than $dz_1 \cdots dz_m$. Hence we can write

$$P(z_1, z_2, \ldots, z_m)\, dz_1\, dz_2 \cdots dz_m$$
$$= \int_{\Gamma} g(\mathbf{a})\, d\alpha_0 \cdots d\alpha_n\, d\beta_0 \cdots d\beta_n, \qquad (4.33)$$

where Γ is the intersection of the subspaces $\mathscr{F}_{z_k}$ for $z_k \in dz_k$. Using (4.30), the *condensed distribution of the roots* of the random algebraic equation $F_n(z, \omega) = 0$ is

$$v^*(S) = (1/n) \int_S P(z_1) \, dz_1. \tag{4.34}$$

As before, let $N_n(B, \omega)$ denote the number of roots of the random algebraic equation lying in the Borel set B of the complex plane Z. Then the condensed distribution $v^*(S)$ can be interpreted as a measure which assigns to the Borel subset S of the complex plane the number, $(1/n)\mathscr{E}\{N_n(S, \omega)\}$.

Hammersley obtained the density of the condensed distribution function in the cases when the coefficients $a_k(\omega)$ are normally distributed complex-valued and real-valued random variables respectively; and in the latter case obtained the result of Kac, as given by (3.3), as a special case. It is of interest to note that the result of Hammersley yields more "information" than Kac's result, in the sense that in addition to giving the average number of real roots it also shows how these roots are distributed on the real line.

V. Some Limit Theorems

In this section we present some results concerning the limiting behavior of the number of roots of a random algebraic equation. We will be concerned with the evaluation of $\lim_{n \to \infty} (1/n)N_n(B, \omega)$, where, as before, $N_n(B, \omega)$ denotes the number of roots of a random algebraic equation of degree n lying in a Borel set B of the complex plane. Consider the random algebraic equation $F_n(z, \omega) = 0$ when the coefficients $a_k(\omega)$ are independent and identically distributed complex-valued random variables. The case where $\mathscr{P}\{a_k(\omega) = 0, \ k = 0, 1, \ldots, n\} = 1$ is not considered. Let α, β, and δ be three arbitrary numbers such that $0 \le \alpha < \beta \le 2\pi$, and $\delta \in (0, 1\}$; and let $B = \{z: \alpha < \arg z < \beta\}$, $C = \{z: 1 - \delta \le |z| \le 1 + \delta\}$.

We now state and prove the following result:

Theorem 5.1. *Let the coefficients $a_k(\omega)$ of the random algebraic equation $F_n(z, \omega) = 0$ be independent and identically distributed complex-valued random variables, and let $\mathscr{E}\{\log^+ |a_k|\} < \infty$.[5] Then, as $n \to \infty$,*

---

[5] $\log^+ |a_k| = \max\{0, \log |a_k|\}$.

(i) $N_n(C, \omega)/n \to 1$
(ii) $N_n(B, \omega)/n \to (\beta - \alpha)/2\pi,$

in probability.

The proof of Theorem 5.1 utilizes the following lemmas.

Lemma 5.1. *The radius of convergence of the random power series*

$$F(z, \omega) = \sum_{k=0}^{\infty} a_k(\omega) z^k, \tag{5.1}$$

is unity almost surely.

Lemma 5.2. *Let ξ_k, $k = 0, 1, \ldots, n$, be arbitrary complex numbers, not all of them zero, and let $N(\alpha, \beta)$ be the number of roots of $\sum_{k=0}^{n} \xi_k z^k$ which lie in the sector $0 \le \alpha < \arg z < \beta$. Then, for $\xi_0 \xi_n \ne 0$,*

$$\left| N(\alpha, \beta) - \frac{(\beta - \alpha)n}{2\pi} \right| < 16 \left[n \log \frac{\sum_{k=0}^{n} |\xi_k|}{[|\xi_0 \xi_n|]^{1/2}} \right]^{1/2}. \tag{5.2}$$

Lemma 5.1 is due to Šparo and Šur [57]; however this result is an immediate consequence of the *zero–one law for random power series* (cf. [2, Theorem 6]). This law can be stated as follows:

Let $F(z, \omega) = \sum_{k=0}^{\infty} a_k(\omega) z^k$ be a random power series, where the moduli of the coefficients $|a_k(\omega)|$ are independent and identically distributed random variables with distribution function $G(\xi)$. Then the radius of convergence $r(\omega)$ of $F(z, \omega)$, where $r(\omega) = \overline{\lim}_{n \to \infty} [|a_n(\omega)|^{1/n}]^{-1}$ is either 0 or 1 almost surely, with

$$r(\omega) = \begin{cases} 0 & \text{a.s.} \quad \Leftrightarrow \int_1^{\infty} \log \xi \, dG(\xi) = \infty \\ 1 & \text{a.s.} \quad \Leftrightarrow \int_1^{\infty} \log \xi \, dG(\xi) > \infty. \end{cases}$$

Lemma 5.2, which is for deterministic polynomials, is due to Erdős and Turán [20].

Proof of theorem 5.7. To simplify the proof we shall assume that $\mathscr{P}\{a_k(\omega) = 0\} = 0$, $k = 0, 1, \ldots, n$. To prove the first part of the theorem, let D_δ denote the open disc of radius δ and with center the origin of the complex plane, let $z_n(\delta, \omega)$ (respectively $z_n'(\delta, \omega)$) be the number of roots of $F_n(z, \omega)$ lying in $D_{1+\delta}$ (respectively, not lying in the closure of $D_{1+\delta}$), and let $w(\delta, \omega)$ denote the number of roots of $F(z, \omega)$ lying in $D_{1-\delta/2}$. Now, an application of Hurwitz's theorem (cf. Hille [32, Vol. II, p. 205]) yields the inequality

$$z_n(\delta, \omega) < w(\delta, \omega) < \infty, \tag{5.3}$$

this inequality holding for all $n \geq n_0(\delta, \omega)$. The random variable $z_n'(\delta, \omega)$ is equal to the number of roots of the polynomial $z^n F_n(z^{-1}, \omega)$ lying in $D_{1/1+\delta}$; moreover, the coefficients of this polynomial are independent and identically distributed random variables as are the coefficients of $F_n(z, \omega)$. Therefore

$$\mathscr{E}\{z_n'(\delta, \omega)\} = \mathscr{E}\{z_n(1 - (1/1 + \delta), \omega)\} = \mathscr{E}\{z_n(\delta/(1 + \delta), \omega)\}. \tag{5.4}$$

Using the equality

$$n - N_n(C, \omega) = z_n(\delta, \omega) + z_n'(\delta, \omega), \tag{5.5}$$

which is satisfied when $a_n(\omega) \neq 0$, we have

$$\mathscr{E}\left\{\frac{n - N_n(C, \omega)}{n}\right\} = \mathscr{E}\left\{\frac{z_n(\delta, \omega) + z_n(\delta/(1 + \delta), \omega)}{n}\right\}. \tag{5.6}$$

The expression within the braces on the right does not exceed 2, since $F_n(z, \omega)$ has n roots at most. Further, using (5.3), the numerator of this expression, for n sufficiently large, does not exceed $w(\delta, \omega) + w(\delta/(1 + \delta), \omega)$. Therefore, passing to the limit, we have that both members of (5.6) tend to zero as $n \to \infty$. Hence $N_n(C, \omega)/n \to 1$ in probability as $n \to \infty$, and this proves the first part of the theorem.

To prove the second part of the theorem we proceed as follows. In view of the restriction $\mathscr{P}\{a_k(\omega) = 0\} = 0$, for any $\varepsilon > 0$ there exists a $\Delta = \Delta(\varepsilon)$ such that $\mathscr{P}\{|a_k(\omega)| > \Delta\} > 1 - \varepsilon$. Therefore

$$\mathscr{P}\{\log(|a_0 a_n|)^{-1/2} < \log 1/\Delta\} > (1 - \varepsilon)^2$$

for all n. We now consider the estimation of $\sum_{k=0}^n |a_k|$. In [2] it is shown that $\mathscr{E}\{\log^+ |a_k|\} < \infty$ if and only if $\sum_{k=0}^\infty \mathscr{P}\{|a_k| \geq (1 + \varepsilon_1)^k\} < \infty$ for all $\varepsilon_1 > 0$. An application of the Borel–Cantelli lemma gives the result

$$|a_k| < (1 + \varepsilon_1)^k,$$

for all $k \geq k_1(\varepsilon_1, \omega)$; hence

$$\sum_{k=0}^n |a_k| < \frac{(1 + \varepsilon_1)^{n+1}}{\varepsilon_1} (1 + C_n(\varepsilon_1, \omega)),$$

where $C_n(\varepsilon_1, \omega) \to 0$ a.s. as $n \to \infty$. Since $\varepsilon_1 > 0$ is arbitrary,

$$(1/n)\log \sum_{k=0}^{n} |a_k| \to 0 \quad \text{a.s.}$$

as $n \to \infty$. Therefore

$$\mathscr{P}\left\{(1/n)\log \frac{\sum_{k=0}^{n} |a_k|}{[|a_0 \, a_n|]^{1/2}} < \varepsilon\right\} > 1 - \varepsilon$$

for all $n \geq n_0(\varepsilon)$. The second assertion now follows from Lemma 5.2.

The interpretation of Theorem 5.1 is clear: The theorem states that, under rather wide conditions, the distribution of the roots of the random algebraic equation $F_n(z, \omega) = 0$ is, in a certain sense, close to the uniform distribution on the circumference of the unit circle (for sufficiently large n) with the center the origin of the complex plane.

Arnold has also proved several limit theorems associated with the roots of random algebraic equations. We state the following typical result.

Theorem 5.2. *Let $a_k(\omega)$, $k = 0, 1, \ldots$ be a sequence of complex-valued random variables, the moduli of which are identically distributed with distribution function $G(\xi) = \mathscr{P}\{|a_k| < \xi\}$. Let*

(i) $\mathscr{P}\{a_k(\omega) = 0\} = G(0^+) = 0$,

(ii) $\mathscr{E}\{\log |a_k(\omega)|\} = \int_0^\infty |\log \xi| \, dG(\xi) < \infty$;

and let $B = \{z: \alpha \leq \arg z \leq \beta, \; 1 - \delta < |z| < 1 + \delta\}$, $0 \leq \alpha < \beta \leq 2\pi$, $\delta \in (0, 1]$.

Then, for the random algebraic equation $F_n(z, \omega) = 0$,

$$\lim_{n \to \infty}(1/n)N_n(B, \omega) = (\beta - \alpha)/2\pi,$$

almost surely and in the rth mean ($r > 0$).

Using results concerning the weak convergence of measures, Theorem 5.2 can be formulated as follows:

Corollary. *Let $z_1(\omega)$, $z_2(\omega)$, $\ldots$, $z_n(\omega)$ be the roots of the random algebraic equation $F_n(z, \omega) = 0$. Then, for all $f(z) \in C(Z)$, where $C(Z)$ is the Banach space of the functions continuous and bounded on Z,*

(i) $\lim_{n \to \infty}(1/n)\sum_{i=1}^{n} f(z_i(\omega)) = (1/2\pi)\int_0^{2\pi} f(e^{i\theta}) \, d\theta$

almost surely and in the rth mean $(r > 0)$; and

(ii) $\lim_{n \to \infty}(1/n) \sum_{i=1}^{n} \mathscr{E}\{f(z_i(\omega))\} = (1/2\pi) \int_0^{2\pi} f(e^{i\theta})\, d\theta.$

Because (i) and (ii) are types of integration formulas, this result should be of interest in numerical analysis.

VI. Random Matrices and Random Algebraic Equations

A. INTRODUCTION

Consider an $n \times n$ matrix $A(\omega) = (a_{ij}(\omega))$, $\omega \in \Omega$. $A(\omega)$ is said to be a *random matrix operator* if $A(\omega)x$ is a random n-vector with values in E_n for every n-vector $x \in E_n$. Equivalently, $A(\omega)$ is a matrix whose n^2 elements $a_{ij}(\omega)$ are random variables, or a random variable with values in E_{n^2}. A random matrix can also be defined as a generalized random variable with values in the Banach algebra of $n \times n$ matrices. The study of random matrices is of interest in its own right, since random matrices form an important class of matrices, and the study of their properties often leads to generalizations of known results in classical matrix theory. In recent years, however, random matrices have found many applications in mathematical statistics, in physics, where they arise as finite-dimensional approximations of random Hamiltonian operators (cf. Mehta [44] and Porter [49]), and in general applied mathematics in connection with systems of random linear equations. Products of random matrices have also led to some interesting classes of limit theorems in abstract probability theory (cf. Grenander [29]).

Consider matrix equations of the form

$$Ax = y, \qquad (6.1)$$

or

$$(A - I)x = y, \qquad (6.2)$$

where $A = (a_{ij})$ is an $n \times n$ matrix, and x and y are n-vectors. If the coefficient matrix A is subject to random error,[6] which is frequently the case in applied problems, the matrix elements a_{ij} can often be written in the form

$$a_{ij}(\omega) = \alpha_{ij} + \varepsilon_{ij}(\omega), \qquad (6.3)$$

---

[6] Cf., for example, Lonseth [42].

where the elements α_{ij} are assumed to be known and the perturbing elements ε_{ij} are random variables. Hence Eqs. (6.1) and (6.2) are random matrix equations.

For applications of systems of random linear equations in economics we refer to Marschak and Andrews [43] and Quandt [52], and for applications to linear programming problems we refer to Babbar [6] and Prékopa [50]. Systems of random linear equations are also encountered in the study of systems of random difference and differential equations. We refer to the dissertation of Nake [45] for a rigorous analysis of systems of random linear equations.

The study of random matrices leads to problems concerning random determinants, random characteristic polynomials, random eigenvalues and their distribution, etc. Some of these problems will be discussed briefly in Section VI.B.

B. EIGENVALUES OF RANDOM MATRICES—RANDOM CHARACTERISTIC POLYNOMIALS

Given an $n \times n$ random matrix $A(\omega) = (a_{ij}(\omega))$ we can, just as in the classical deterministic case, introduce the associated *random determinant*[7]

$$|A(\omega)| = \begin{vmatrix} a_{11}(\omega) & \cdots & a_{1n}(\omega) \\ a_{21}(\omega) & \cdots & a_{2n}(\omega) \\ \vdots & \vdots & \vdots \\ a_{n1}(\omega) & \cdots & a_{nn}(\omega) \end{vmatrix}. \tag{6.4}$$

The value of $|A(\omega)|$ is a random variable. We can now consider the *homogeneous random eigenvalue equation*

$$(A(\omega) - \lambda I)x = 0, \tag{6.5}$$

which has a nontrivial solution if and only if λ satisfies the *characteristic equation* of $A(\omega)$ obtained by equating the determinant of the random matrix $A(\omega) - \lambda I$ to zero. Such a λ is called a *characteristic* or *eigenvalue* of $A(\omega)$; and clearly these λ will depend on ω, and will, in general, be random variables. The characteristic polynomial associated with an $n \times n$ matrix will be a random algebraic polynomial of degree n, and the solutions or roots of the random algebraic equations will be the eigenvalues of $A(\omega)$.

---

[7] For a discussion of random determinants and their properties we refer to the papers of Bellman [7], Fortet [25], Nyquist *et al.* [47], and Prékopa [51].

In studies concerned with determining the distribution of the eigenvalues of a random matrix, the distribution of the roots of the associated
random characteristic polynomial is seldom if ever investigated, and
other methods must be utilized. One of the main reasons for not working
with random algebraic equations can be demonstrated by considering a
2×2 random matrix $B(\omega)$. In this case the associated random characteristic equation is of the form

$$\lambda^2 + a_1(\omega)\lambda + a_0(\omega) = 0, \tag{6.6}$$

where

$$\begin{aligned}
a_1(\omega) &= -(b_{11}(\omega) + b_{22}(\omega)), \\
a_0(\omega) &= b_{11}(\omega)b_{22}(\omega) - b_{12}(\omega)b_{21}(\omega).
\end{aligned} \tag{6.7}$$

Hence, given the joint distribution of the elements of $B(\omega)$, it is first
necessary to determine the distribution of the coefficients $a_1(\omega)$ and
$a_2(\omega)$ as given by (6.7). Then, given the joint distribution of $a_1(\omega)$ and
$a_2(\omega)$ it remains to determine the joint distribution of the random
eigenvalues $\lambda_1(\omega)$ and $\lambda_2(\omega)$.

For an $n \times n$ random matrix $B(\omega) = (b_{ij}(\omega))$ the relationship between
the matrix elements $b_{ij}(\omega)$ and the coefficients $a_k(\omega)$ of the characteristic
polynomial $F_n(\lambda, \omega)$ is given by Newton's formula (cf. Gantmacher
[26b, p. 87]), which expresses the coefficients $a_k(\omega)$ in terms of the traces
$t_1(\omega)$, $t_2(\omega)$, $\ldots$, $t_n(\omega)$ of the matrices $B(\omega)$, $B^2(\omega)$, $\ldots$, $B^n(\omega)$. We refer
also to the methods of Faddeev and Krylov (cf. [26b, pp. 87–89, pp.
202–214]) for the computation of the coefficients of the characteristic
polynomial. Hence, in principle, a transformation of the above type,
followed by a transformation of the type discussed in Section IV.C,
enables us to compute the joint density of the random eigenvalues.

Another reason that the distribution of the roots of random algebraic
equations has not been utilized in the development of the spectral theory
of random matrices is due to the fact that most of the studies in mathematical statistics and statistical physics which lead to random matrices
have been concerned with the asymptotic distribution of the random
eigenvalues (cf. Arnold [5], Grenander [29], Mehta [44], and Porter [49]).

In conclusion, a systematic study of the distribution of the roots of
random characteristic polynomials would be of great importance, and of
great value in the study of systems of random algebraic, difference, and
differential equations, stochastic stability theory (cf. Saaty [55, pp.
418–419]), and eigenvalue problems for random integral equations
(cf. Bharucha-Reid and Arnold [11]).

ACKNOWLEDGMENTS

The author would like to thank Professor L. Arnold and Mr. D. Kannan for
reading preliminary versions of this manuscript, and for their very helpful comments.

REFERENCES

1. Anderson, T. W., "Introduction to Multivariate Statistical Analysis," Wiley
   (Interscience), New York, 1958.
2. Arnold, L., Über die Konvergenz einer zufälligen Potenzreihe, *J. Reine Angew.
   Math.* **222** (1966), 79–112.
3. Arnold, L., Über die Nullstellenverteilung zufälliger Polynome, *Math. Z.* **92**
   (1966), 12–18.
4. Arnold, L., Random power series, *Statistical Laboratory Publications* (SLP-1),
   Michigan State Univ., East Lansing, Michigan, 1966.
5. Arnold, L., On the asymptotic distribution of the eigenvalues of random matrices,
   J. Math. Anal. Appl. **20** (1967), 262–268.
6. Babbar, M. M., Distributions of solutions of a set of linear equations (with an
   application to linear programming), *J. Amer. Statist. Assoc.* **50** (1955), 854–869.
7. Bellman, R., A note on the mean value of random determinants, *Quart. Appl.
   Math.* **13** (1955), 322–324.
8. Belyaev, Yu. K., Analytic random processes, (Russian), *Teor. Verojatnost. i
   Primenen.* **4** (1959), 437–444.
9. Berry, A. C., The accuracy of the Gaussian approximation to the sum of independent variates, *Trans. Amer. Math. Soc.* **49** (1941), 122–136.
10. Bharucha-Reid, A. T., On the theory of random equations, *Proc. Symp. Appl.
    Math.* **16**, pp. 40–69, Amer. Math. Soc., Providence, Rhode Island, 1964.
11. Bharucha-Reid, A. T. and Arnold, L., On Fredholm integral equations with
    random degenerate kernels, *Zastos. Mat.* **10** (Steinhaus Jubilee Volume) (1969),
    85–90.
12. Bloch, A. and Pólya, G., On the roots of certain algebraic equations, *Proc.
    London Math. Soc.* **33** (1932), 102–114.
13. Cramér, H. and Leadbetter, M. R., "Stationary and Related Stochastic Pro-
    cesses," Wiley (Interscience), New York, 1967.
14. Das, M., The average number of real zeros of a random trigonometric poly-
    nomial, *Proc. Cambridge Philos. Soc.* **64** (1968), 721–729.
15. Dunnage, J. E. A., The number of real zeros of a random trigonometric poly-
    nomial, *Proc. London Math. Soc.* **16** (1966), 53–84.
16. Dunnage, J. E. A., The number of real zeros of a class of random algebraic
    polynomials, *Proc. London Math. Soc.* **18** (1968), 439–460.
17. Durand, E., "Solutions numériques des équations algébriques." Vol. I. "Équa-
    tions du type $F(x) = 0$. Racines d'un polynôme," Masson, Paris, 1960.
18. Erdélyi, A. (ed.), "Tables of Integral Transforms," Vol. I, McGraw-Hill, New
    York, 1954.
19. Erdős, P. and Offord, A. C., On the number of real roots of a random algebraic
    equation, *Proc. London Math. Soc.* **6** (1956), 139–160.

20. Erdős, P. and Turán, P., On the distribution of roots of polynomials, *Ann. of Math.* **51** (1950), 105–119.
21. Esseen, C. G., Fourier analysis of distribution functions, *Acta Math.* **77** (1945), 1–125.
22. Evans, E. A., On the number of real roots of a random algebraic equation, *Proc. London Math. Soc.* **15** (1965), 731–749.
23. Fairley, W., Roots of random polynomials, Dissertation, Harvard Univ., Cambridge, Massachusetts, 1968.
24. Finney, D. J., "Statistical Method in Biological Assay," Griffen, London, 1952.
25. Fortet, R., Random determinants, *J. Res. Nat. Bur. Standards Sect. B* **17** (1951), 465–470.
26. Gaede, K. -W., Die Verteilung von gewissen reellen Wurzeln der Gleichung *n*-ten Grades mit reellen Zufallskoeffizienten, *Monatsh. Math.* **63** (1959), 359–367.
26a. Gaede, K. -W., Zur Verteilung der Wurzeln zufälliger algebraischer Gleichungen, *Math. Nachr.* **21** (1960), 81–107.
26b. Gantmacher, F. R., "The Theory of Matrices," Vol. 1, Chelsea, New York, 1959.
27. Geary, R. C., The frequency distribution of the quotient of two normal variables, *J. Roy. Statist. Soc. Ser. A* **93** (1930), 442–446.
28. Girshick, M. A., Note on the distribution of roots of a polynomial with random complex coefficients, *Ann. Math. Statist.* **13** (1942), 235–238.
29. Grenander, U., "Probabilities on Algebraic Structures," Wiley (Interscience), New York, 1963.
30. Hamblen, J. W., Distribution of roots of quadratic equations with random coefficients, *Ann. Math. Statist.* **27** (1956), 1136–1143.
31. Hammersley, J., The zeros of a random polynomial, *Proc. Third Berkeley Symp. Math. Statist. Prob., 1955* **2**, 89–111, 1956.
32. Hille, E., "Analytic Function Theory," Vols. I and II, Ginn, Boston, 1959 and 1962.
33. Kabe, D. G., On the distribution of the roots of some random algebraic equations. Unpublished manuscript, 1964.
34. Kac, M., On the average number of real roots of a random algebraic equation, *Bull. Amer. Math. Soc.* **49** (1943), 314–320, 938.
35. Kac, M., On the average number of real roots of a random algebraic equation (II), *Proc. London Math. Soc.* **50** (1948), 390–408.
36. Kac, M., "Probability and Related Topics in Physical Sciences," Wiley (Interscience), New York, 1959.
37. Kahane, J.-P., "Some Random Series of Functions," Heath, Lexington, Massachusetts, 1968.
38. Littlewood, J. E. and Offord, A. C., On the number of real roots of a random algebraic equation, *J. London Math. Soc.* **13** (1938), 288–295.
39. Littlewood, J. E. and Offord, A. C., On the number of real roots of a random algebraic equation (II) *Proc. Cambridge Philos. Soc.* **35** (1939), 133–148.
40. Littlewood, J. E. and Offord, A. C., On the number of real roots of a random algebraic equation (III), *Mat. Sb.* **12** (1943), 277–286.
41. Logan, B. F. and Shepp, L. A., Real zeros of random polynomials, *Proc. London Math. Soc.* **18** (1968), 29–35.
41a. Logan, B. F. and Shepp, L. A., Real zeros of random polynomials. II, *Proc. London Math. Soc.* **19** (1968), 308–314.

42. Lonseth, A. T., The propagation of error in linear problems, *Trans. Amer. Math. Soc.* **62** (1947), 193–212.

43. Marschak, J. and W. H. Andrews, Random simultaneous equations and the theory of production, *Econometrica* **12** (1944), 143–205.

44. Mehta, M. L., "Random Matrices and the Statistical Theory of Energy Levels," Academic Press, New York, 1967.

45. Nake, F., Über die Anzahl der reellen Lösungen zufälliger Gleichungssysteme, Dissertation, Technische Hochschule Stuttgart, 1967.

46. Neveu, J., "Bases mathématique du calcul des probabilitiés," Masson, Paris, 1964.

47. Nyquist, H., Rice, S. O., and Riordan, J., The distribution of random determinants, *Quart. Appl. Math.* **12** (1954), 97–104.

48. Offord, A. C. An inequality for sums of independent random variables, *Proc. London Math. Soc.* **48** (1945), 467–477.

49. Porter, C. E. (ed.), "Statistical Theories of Spectra: Fluctuations," Academic Press, New York, 1965.

50. Prékopa, A., On the probability distribution of the optimum of a random linear problem, *J. Soc. Indust. Appl. Math. Ser. A Control* **4** (1966), 211–222.

51. Prékopa, A., On random determinants I, *Studia Sci. Math. Hungar.* **2** (1967), 125–132.

52. Quandt, R. E., Probabilistic errors in the Leontief system, *Naval Res. Logist. Quart.* **5** (1958), 155–170.

53. Rice, S. O., Mathematical analysis of random noise, *Bell System Tech. J.* **23** (1944), 282–332 and **24** (1945), 45–156.

54. Rudin, W. "Real and Complex Analysis," McGraw-Hill, New York, 1966.

55. Saaty, T. L., "Modern Nonlinear Equations," McGraw-Hill, New York, 1967.

56. Samal, G., On the number of real roots of a random algebraic equation, *Proc. Cambridge Philos. Soc.* **58** (1962), 433–442.

57. Šparo, D. I. and Šur, M. G., On the distribution of roots of random polynomials (Russian), *Vestnik Moskov. Univ. Ser. I Mat. Meh.* (1962), 40–43.

58. Stevens, D. C., The average and variance of the number of real zeros of random functions, Dissertation, New York Univ., New York, New York, 1965.

59. Stevens, D. C., The average number of real zeros of a random polynomial, *Comm. Pure Appl. Math.* **22** (1969), 457–477.

60. Whittaker, E. T. and Watson, G. N., "A Course of Modern Analysis," 4th ed., Cambridge Univ. Press, London and New York, 1958.

61. Wilks, S. S., "Mathematical Statistics," 2nd ed., Princeton Univ. Press, Princeton, New Jersey, 1946.

62. Zinger, M. Ja., On the roots of trigonometric polynomials, (Russian), *Dokl. Akad. Nauk SSSR* **161** (1965), 1263–1266.

AXIOMATIC QUANTUM MECHANICS AND GENERALIZED PROBABILITY THEORY

*Stanley Gudder**

DEPARTMENT OF MATHEMATICS
UNIVERSITY OF DENVER
DENVER, COLORADO

I. Introduction

Modern quantum mechanics is in many respects immensely successful in its description of natural phenomena. The modern theory has produced results which numerically agree with experiment to previously

* Some of the research for this article was done at the Mathematics Research Center, Madison, Wisconsin. During the rest of the preparation, the author was partially supported by N.S.F. Grant GP-6764.

unheard of accuracies. The theory has further led to predictions concerning undiscovered phenomena. It has shown physicists "where to look," for example, for elementary particles and successfully predicted properties of these particles. For these reasons it is generally agreed that this is an essentially "correct" theory in the sense that, although later theories may differ radically from the present one, they will contain the contemporary theories as approximations.

Nothwithstanding its many successes, the present quantum theory also suffers from important inadequacies. First of all, the body of phenomena encompassed by the theory is not as inclusive as might be hoped. Also the theory is deficient, in essential places, in mathematical rigor (cf. [46]). For example, certain perturbation expansions are derived which unfortunately do not converge although the first few terms give numbers agreeing closely with experiment. Furthermore, divergent integrals appear which are ignored in a "renormalization" process (cf. [11]). Although it has not been particularly true in the past, many theoretical physicists today feel that mathematical rigor and physical understanding go hand in hand; and that mathematically inadmissible operations and objects, even though they may have some vague physical significance, cannot be substituted for well-defined quantities and rigorously logical manipulations. For example, in the past physicists have happily manipulated unbounded operators in a purely formal algebraic manner without regard to the domains of these operators. Furthermore, in some cases they have carried through arguments in infinite dimensional spaces using finite-dimensional techniques, and assumed them to go through in the more general situation; even if the arguments did not, the fault was with the mathematics and not with the physics. Today however, the theoretical physicist is concerned with these mathematical details and most feel that this rigor must be adhered to if valid and meaningful results are to be obtained (cf. [45, 54]).

One of the important groups of physicists and mathematicians which has attempted to place quantum physics on a firm mathematical foundation is the "axiomatic school." The main objective of this school is to state a minimum number of simple, physically significant axioms, which ideally can be tested experimentally, and from which previously known phenomena can be derived and heretofore unknown results predicted. At the foundation of quantum mechanics there seem to be three basic concepts: *states*, *propositions*, and *observables*. Using these concepts, essentially two approaches to axiomatic quantum mechanics have evolved. The first approach, due to Jordan *et al.* [23] and von Neumann [58], takes the observables, while the second approach, due to Birkhoff [2] and Birkhoff and von Neumann [3], takes the propositions as the basic axiomatic elements. These two approaches have matured into distinct axiomatic schools of thought. The first school was developed by Segal (see [27, 46, 49, 50, 51]) and forms the basis of the quantum field theories of Haag and Wightman (see [26, 39, 49, 54]). The second approach has been studied to some extent by Bodiou [4]; and Mackey [29, 35]; and more recently by Jauch, Piron, and others (see [8, 18, 19, 21, 36, 41, 42, 50, 62, 63]). This article is mainly concerned with the *second* approach, although some attempt will be made to compare it with the first.

The first approach has the attribute of taking full advantage of the well-developed theories of Banach spaces, B^*-algebras, and W^*-algebras. However, the second has the advantage of giving a direct generalization of the probability theory of Kolmogorov, as pointed out by Bodiou [4] and Varadarajan [57]. The first approach also has a close relationship to probability theory, and considerable work has been done on its probabilistic aspects [47, 48, 53, 55, 56]. However, their methods are not as obviously and transparently related to Kolmogorov's probability theory as the second approach. This is due mainly to the fact that the random variables are taken as the axiomatic elements in the first, while in the second the events are chosen as the undefined elements.

It is almost unanimously agreed that quantum mechanics is, by its very intrinsic nature, a probabilistic theory. (We shall give a short discussion later about some dissenters to this viewpoint.) This is in contrast to statistical mechanics in which statistical methods are admittedly used only as an approximation to obtain macroscopic data while no claim is made for a microscopic description. Because of this nature of quantum physics it seems advisable to take advantage of the powerful machinery of modern probability theory whenever possible; in fact, physicists are doing this today. For example, the Feynman integral of quantum mechanics is similar in form to stochastic integrals which have been studied by probabilists for many years (cf. [38]).

The main purpose of this article is to develop an axiomatic model for quantum mechanics, to give some of the consequences of this model, and to discuss the generalized probability theory that results.

II. Historical Background

Quantum mechanics was born in a furnace. Although this statement may appear ludicrous it is true that the theory of quantum mechanics was developed to remedy discrepancies between an experimentally observed phenomenon in heat radiation and its theoretical explanation. It is well known that heated objects emit electromagnetic radiation of different wavelengths. When the temperature is comparatively low, say at the boiling point of water, the predominant wavelengths of the emitted radiation are rather long and the radiation is mostly in the form of heat. When the temperature rises to 600°C (about that of an electric range) the radiation starts affecting the retina of the eye and we see a faint red light. At 2000°C (an electric light bulb) a bright white light is seen: This light is a mixture of all the wavelengths of the visible radiation spectrum from red to violet. At still higher temperatures a considerable amount of invisible ultraviolet radiation is emitted. Now suppose we have a furnace which can be heated to arbitrary temperatures. By opening a small door in the furnace we can measure the energy of the emitted radiation at different wavelengths or frequencies. It was found experimentally that the energy per unit volume E of the emitted radiation as a function of temperature T and frequency v is

$$E = 4\pi h v^3 \exp\{-hv/kT\}/c^3[1 - \exp\{-hv/kT\}], \qquad (2.1)$$

where c is the velocity of light, k is Boltzmann's constant, and h is a constant which we shall explain later. Thus the energy per unit volume is concentrated at higher and higher frequencies as the temperature rises.

Using classical mechanics, in the late 1800's attempts were made to derive Eq. (2.1) theoretically. One of these attempts was the *Rayleigh–Jeans radiation law* derived using classical statistical mechanics. Suppose we have radiation emitted through a small door of a cubical furnace of length L. The radiation emitted through the open furnace door consists of waves of all frequencies, although of course, certain frequencies are predominant. The energy of the radiation $e(v)$ at different frequencies is found by determining the number of waves $n(v)$ with frequency v and multiplying by the energy associated with these waves. According to the equipartition principle each wave should carry the same amount of energy e_0 on the average, and thus $e(v) = n(v)e_0$. To find $n(v)$ the waves are represented mathematically by plane waves $\exp[2\pi i(\mathbf{k} \cdot \mathbf{r} - vt)]$ where $\mathbf{k}$ is a vector in the direction of the wave propagation with components $(n_1/L, n_2/L, n_3/L)$, where n_1, n_2, n_3 are integers, and $|\mathbf{k}| = v/c$. The integers appear since we must have a whole number of wavelengths in the furnace cavity. Let us now find the number of possible waves $n(v) \, \Delta v$ with frequency between v and $v + \Delta v$. Draw a three-dimensional lattice with dots at each point with integer coordinates and draw a spherical shell whose inner and outer radii are k and $k + \Delta k$, respectively. Then the number of waves whose wavenumber k is between k and $k + \Delta k$ is the number of dots in this shell. This number is approximately the volume of the shell divided by the volume surrounding each dot and thus equals

$$[(k + \Delta k)^3 - k^3]4\pi L^3/3 \cong 4\pi L^3 k^2 \, \Delta k.$$

Thus $n(v) \, \Delta v = 4\pi L^3 v^2 \, \Delta v/c^3$.

We must now find the average energy e_0 of a wave with frequency v. According to statistical mechanics if we have a large number of objects about which we have a maximum amount of ignorance, then the probability density for the energy of an object of our system is given by the Gibb's density $c_0 \exp(-e/KT)$. We may find the constant c_0 using the normalization

$$1 = c_0 \int_0^\infty \exp(-e/KT) \, de = c_0 \, KT.$$

Thus

$$KTe_0 = \int_0^\infty e \exp(-e/KT) \, de = 1,$$

and $e_0 = (KT)^{-1}$. Hence the energy per unit volume of a wave with frequency v according to this method is

$$E = 4\pi KTv^2/c^3. \tag{2.2}$$

Comparing Eqs. (2.2) and (2.1) we see that the Rayleigh–Jeans law agrees with experiment only for low v. For high v Eq. (2.2) leads to the ridiculous conclusion, called the ultraviolet catastrophy, which says that the total energy per unit volume due to all frequencies is infinite.

The correct radiation law Eq. (2.1) was derived theoretically by Max Planck and was presented on December 14, 1900 at a meeting of the German Physical Society. It was here that the quantum of energy was first described. Planck's idea was so unusual and so grotesque at that time, that he himself could hardly believe it. In fact, as we shall see, this idea was an accident which turned out differently than Planck had intended. It forms, however, one of the cornerstones of modern physics. Planck suggested that instead of allowing the energy of the radiation waves to have arbitrary values we make the following postulate:

The energy of electromagnetic waves can exist only in the form of certain discrete packages, or quanta, the energy content of each package being directly proportional to its frequency.

Using this postulate Planck then derived his radiation law. Suppose we have a wave with frequency v. It may have only a countable number of energy values $e(n) = nhv$, $n = 0, 1, 2, \ldots$, where h is constant. Using the Gibb's probability density, the probability that the wave has energy $e(n)$ is

$$p_n = c_0 \exp(-e(n)/KT).$$

Again to find c_0 we use the normalization

$$1 = \sum p_n = c_0 \sum \exp(-nhv/KT) = c_0[1 - \exp(-hv/KT]^{-1}.$$

The average energy of a wave with frequency v becomes:

$$e_v = \sum e(n)p_n = hv[1 - \exp(-hv/KT)] \sum n \exp(-nhv/KT).$$

Using the identity

$$\sum ne^{-nx} = e^{-x}(1 - e^{-x})^2,$$

we have

$$e_v = hv \exp(-hv/KT)[1 - \exp(-hv/KT)]^{-1},$$

which, unlike the classical case, is a function of v. Multiplying this by $n(v)/L^3$ we get Eq. (2.1). In order to get experimental agreement we must have $h = 6.62 \times 10^{-27}$ erg sec.

This last constant h is called Planck's constant. Planck had originally intended h to be nothing more than a mathematical catalyst and had planned to let h go to zero in the final radiation law. In this case the radiation law would tend to the incorrect Rayleigh–Jeans law, so h had to stay constant. Note that since h is very small the high-energy values are almost continuous.

Planck's basic postulate was further substantiated by other phenomena such as Einstein's photoelectric effect and the Compton effect. However, one of its most important applications appeared in Bohr's quantum orbits. In 1911 Ernest Rutherford, at the University of Manchester, using scattering experiments of α-particles through gold foil, discovered the model for the elements which is essentially accepted today. Rutherford came to the conclusion that an atom must consist of a very small positively charged nucleus surrounded by enough electrons (negatively charged particles) to balance the nuclear charge. These experimental findings again contradicted classical theory. The electrons could not be stationary since then they would be attracted to the nucleus causing the atom to collapse, so the electron must be orbiting the nucleus. But by classical electrodynamics, a charged particle which is accelerating must radiate energy and one can calculate, using classical theory, that the electron should radiate all of its energy away within 10^{-8} sec. Having lost all their energy, atomic electrons must fall into the nucleus and the atom ceases to exist! (It is interesting to note that this phenomenon actually occurs in planetary motion where the planets radiate gravitational waves. However, since Newton's constant is very small it will take trillions of years for Earth to lose all its energy.)

There were other phenomena concerning atoms which classical physics could not explain, the most important of these being the light spectrum emitted by heated elements. If an element is heated to a high temperature it will emit light consisting of a sequence of sharp spectral lines which show up on a photographic plate after the light is refracted by a prism. It was found in the 1800's that the spectral lines emitted by heated hydrogen had frequencies which could be labeled by two integers m, n and

$$v_{m, n} = R(n^{-2} - m^{-2}), \qquad n < m, \tag{2.3}$$

where $R = 3.289 \times 10^{15}$ sec^{-1}. In 1913 Bohr gave a theoretical explanation for this. He reasoned that the light emitted by an excited hydrogen atom was due to a loss of energy by its single orbital electron. Since the energy carried by the light must, according to Planck's postulate, have only discrete values, the energy of the orbital electron must also have discrete energy values E_1, E_2, ... When an electron jumps from an energy level E_m to a lower energy level E_n the energy of the emitted light is $hv_{m, n} = E_m - E_n$ and hence $v_{m, n} = (E_m - E_n)/h$. Bohr stated the following postulates:

(1) The hydrogen atom has one electron which moves in a circular orbit and the electron's energy remains constant so long as it is in the same orbit.

(2) Only a discrete number of orbits are possible. In the allowed orbits, the electron must have angular momentum equal to a multiple of $h/2\pi = \hbar$.

(3) When an electron shifts from one possible orbit to another, the energy difference $E_m - E_n$ is transformed into radiation of frequency $(E_m - E_n)/h$.

We now show that the spectroscopic data in Eq. (2.3) follows from these three postulates. Suppose the electron is in the nth possible orbit with radius r_n and let μ, e be the known mass and charge of the electron.

Since from (1) the electron moves in a circular orbit, it follows by Newton's second law that $\mu v_n{}^2/r_n{}^2 = e^2/r_n{}^2$, where v_n is the speed of the electron. Thus $v_n = e(\mu r_n)^{-1/2}$. From (2) we have $\mu v_n r_n = nh$. Solving these two equations for r_n we get $r_n = n^2 h^2/\mu e^2$. In particular $r_1 = h^2/\mu e^2 \cong 10^{-8}$ cm which agrees with experiment. To find the total energy E_n, we know that E_n is the sum of the kinetic energy K_n and the potential energy V_n. Thus

$$E_n = K_n + V_n = \mu v_n{}^2/2 - e^2/r_n = -2\pi^2 e^2 \mu/n^2 h^2. \qquad (2.4)$$

Hence the frequencies of the emitted light should be

$$v_{n,\,m} = (E_m - E_n)/h = 2\pi^2 e^4 \mu h^{-3}(n^{-2} - m^{-2}).$$

Now from the known values of e, μ, h one finds that $2\pi^2 e^4 \mu h^{-3} = R$ and hence (2.3) is derived.

Electromagnetic radiation is naturally associated with a wave-type motion. However, in certain experiments such as the photoelectric effect and the Compton effect, radiation acts like a stream of particles or photons. In 1925 de Broglie postulated that particles should also have this dual role and thus, for example, electrons should have wave-like properties. He asserted that the hydrogen orbital electron in its nth quantum orbit should have associated with it n waves which "fit" in the orbit. Thus the shortest wavelength of the n waves must be $\lambda_n = 2\pi r_n/n$. Substituting for r_n and v_n according to our previous formulas we get $\lambda_n = h/\mu v_n$. De Broglie then concluded that an electron should have associated with it a wave of wavelength $h/\mu v$. This was later substantiated experimentally when it was shown that electron beams gave diffraction patterns similar to light waves and that these patterns corresponded to waves of wavelength $h/\mu v$. Now electromagnetic waves in vacuum satisfy a differential equation of the form

$$c^{-2} \frac{\partial^2 E}{\partial t^2} = \nabla^2 E = \frac{\partial^2 E}{\partial x^2} + \frac{\partial^2 E}{\partial y^2} + \frac{\partial^2 E}{\partial z^2},$$

so one might expect that de Broglie waves also satisfy a similar differential equation. In 1926 Schrödinger discovered the equation which governs de Broglie waves in vacuum:

$$\partial \psi/\partial t = -(\hbar^2/2\mu)\, \nabla^2 \psi.$$

This equation was later generalized by Dirac to include relativistic effects. For the hydrogen atom the equation becomes

$$\partial \psi/\partial t = -(\hbar^2/2\mu)\, \nabla^2 \psi - (e^2/r)\psi.$$

Schrödinger postulated that the eigenvalues E of the equation

$$-(\hbar^2/2\mu)\,\nabla^2\psi - (e^2/r)\psi = E\psi$$

with suitable boundary conditions give the allowed values for the electrons energy. These eigenvalues turn out to be just the values given in (2.4) and thus Schrödinger's postulate leads to the same results as Bohr's postulates.

In 1927 Heisenberg announced his famous uncertainty principle. This principle states that the position and momentum of a physical object cannot be measured with arbitrary precision at the same time. In fact, if Δp and Δq are the errors made in a momentum and coordinate measurement made at the same time, then $\Delta p\,\Delta q \geq h$. This principle was substantiated by experiment. As a simple example suppose we have an electron moving along the x-axis and we want to measure its position and momentum. To find the electron's position we must "see" it and hence must bounce a light ray off of it. If the light ray has wavelength λ, then $\Delta x \cong \lambda$. The light ray will then impart an additional momentum $h\nu/c = h/\lambda$ to the electron and thus $\Delta p \cong h/\lambda$ and $\Delta x\,\Delta p \cong h$. This gives an example of "interfering" measurements.

Because of the smallness of h, the uncertainty principle is insignificant so far as everyday objects are concerned. For example, for an object weighing 1 g we have $\Delta x\,\Delta v \geq 10^{-27}$ cm^2 sec^{-1}, so if we knew its position within 10^{-10} cm we could still find the velocity to within 10^{-17} cm sec^{-1}. On the other hand, the mass of an electron is approximately 10^{-27} g so $\Delta x\,\Delta v \geq 1$ cm^2 sec^{-1}. If we know that an electron is in a particular atom, then $\Delta x \cong 10^{-8}$ cm and therefore $\Delta v \geq 10^8$ cm sec^{-1}. This gives an uncertainty in kinetic energy $K = \frac{1}{2}\mu v^2$ of $\Delta K = \mu v\,\Delta v \cong 10^{-27} \cdot 10^8 \cdot 10^8 \cong 10$ eV, which is comparable to the total binding energy of an electron in the atom. Conversely, if we know an electron is in an atom, then we know $\Delta K \leq 10$ eV and thus $\Delta x \geq 10^{-8}$ cm. It is therefore ridiculous to think of electrons as moving around the nucleus in a definite orbit since the width of this orbit would be about the same size as the atom itself. One therefore, usually thinks of electrons as being smeared in a "cloud" around the nucleus.

Thinking in terms of the uncertainty principle, it is usually concluded that quantum physics is intrinsically a probabilistic theory. One cannot give arbitrarily precise values for all measurements on a physical system but can only predict these values probabilistically. This statistical interpretation was also given by Born. He proposed that the de Broglie waves were not waves in a physical sense, but were probability waves, in the sense that the de Broglie wave $\psi(x, y, z)$ is a complex function which determines the probability that the system is at the point (x, y, z). More precisely, the probability that the system is in a set $\Lambda \subset R^3$ is

$$\int_\Lambda |\psi(x, y, z)|^2 \, dx \, dy \, dz,$$

assuming that

$$\int_{R^3} |\psi(x, y, z)|^2 \, dx \, dy \, dz = 1.$$

In summary, we have tried in this section to give some of the important highlights of quantum mechanics which developed between 1900 and 1930. One can see that quantum mechanics consisted of a collection of vaguely related theories held together by a thin thread. The theories were stated in terms of certain "ad hoc" proposals such as Planck's energy quanta, Bohr's quantization rules, de Broglie–Schrödinger waves, and Heisenberg's uncertainty principle. The only justification that could be given for these postulates is that they led to results which agreed with experiment. However, they had the disadvantage of applying only to very specific phenomena and could not be easily generalized, without adding great complications, to more sophisticated phenomena. Starting about 1930, von Neumann, Wigner, Jordan, and others attempted to give an axiomatic formulation of quantum mechanics which would excompass all of these theories. It is the modern developments of these attempts which we shall consider here.

III. Classical Mechanics

Before we can attempt a formulation of quantum mechanics it is important to understand some of the basic principles of classical mechanics. In this section we shall follow [29, pp. 1–8] fairly closely. A mechanical system has N *degrees of freedom* if its configuration can be completely described by N continuous functions called *coordinate functions* or *coordinates*, but not by $N - 1$ continuous functions, in a finite-dimensional Euclidean space. For example, a particle moving around a circle or constrained to move in a straight line has one degree of freedom, and a particle moving on the surface of a sphere in three-dimensional space R^3 has two degrees of freedom. If we have n interacting particles in R^3, we have $3n$ degrees of freedom and thus the configuration of the system is completely described by coordinate functions $x_1(t)$, $y_1(t)$, $z_1(t)$, ..., $x_n(t)$, $y_n(t)$, $z_n(t)$ in R^{3n} space. We will always assume that the coordinate functions are twice differentiable with respect to the time t. The fundamental principle of classical mechanics is:

(C) *the coordinate functions at time t_2 are determined by the coordinate functions and their first time derivative at time t_1 if $t_2 \geq t_1$.*

This is just a generalization of Newton's second law. For example, if a particle is contrained to move in the x-direction Newton's second law states that

$$m(d^2 x/dt^2) = f(t, x, (dx/dt)).$$

If f is sufficiently nice (e.g., f, $(\partial f/\partial x)$, $(\partial f/\partial y)$, where $y = dx/dt$, are continuous functions) then there is one and only one solution of the differential equation satisfying any initial conditions.

We call the time derivatives of the coordinates *velocity functions* or *velocities*. If a mechanical system has N degrees of freedom with coordinates $q_1, \ldots, q_N$ and velocities $v_1, \ldots, v_N$, the *state* $s(t_0)$ of the system at time t_0 is

$$s(t_0) = (q_1(t_0), \ldots, q_N(t_0), v_1(t_0), \ldots, v_N(t_0)).$$

Thus the state of the system is given by $2N$ numbers and hence the set of states S may be thought of as a subset of $2N$ dimensional space. If we know the state at any time we have a complete description of the system at that time. The important property of the states is that by (C) if we know the state at one time we may determine it at all later times. For $s \in S$ and $t \geq 0$, let $U_t(s)$ denote the state at time t when the state at time $t = 0$ is s. Now by (C) U_t is well defined and of course $U_0 = I$, the identity map. Thus for each $t \geq 0$, U_t is a transformation of S into S. Now $U_{t_1}(U_{t_2}(s))$ is the state t_1 time units after the state was $U_{t_2}(s)$ and $U_{t_2}(s)$ the state t_2 time units after it was s. Thus $U_{t_1}(U_{t_2}(s))$ is the state $t_1 + t_2$ time units after it was s and hence for $t_1, t_2 \geq 0$ we have:

$$U_{t_1 + t_2} = U_{t_1} U_{t_2}. \tag{3.1}$$

A collection of transformations $\{U_t, t \in (0, \infty)\}$ satisfying (3.1) is called a *one-parameter semigroup* of transformations. A mechanical system is *reversible* if U_t is a one–one onto transformation. For simplicity we will only consider reversible systems. In this case $(U_t)^{-1}$ exists and we define $U_{-t} = (U_t)^{-1}$. A collection of transformations for which (3.1) holds for all $t_1, t_2 \in (-\infty, \infty)$ is called a *one-parameter group* of transformations. If $s_0 \in S$, an *orbit* in S is a set $\{U_t(s_0) : t \in (-\infty, \infty)\}$. The following lemma is easily checked.

Lemma 3.1. *If a mechanical system is reversible, then U_t is a one-parameter group and each $s \in S$ lies on one and only one orbit.*

We call $U = \{U_t, t \in (-\infty, \infty)\}$, the *dynamical group* of the system. If we know the dynamical group for a mechanical system we have a complete description of the system at all times. However, we usually do not know the dynamical group U *a priori* but only the tangent vectors along its orbits (usually given by Newton's second law) and we must derive U from these. That is, we are given $2N$ functions A_i^0, A_i, $i = 1, \ldots, N$ of $t, q_1, \ldots, q_N, v_1, \ldots, v_N$ such that

$$dq_i/dt = A_i^0(t, q_1, \ldots, q_N, v_1, \ldots, v_N),$$

$$dv_i/dt = A_i(t, q_1, \ldots, q_N, v_1, \ldots, v_N), \qquad i = 1, \ldots, N.$$

If the A_i^0, A_i are continuous and differentiable with respect to $q_1, \ldots, v_N$ we say that U is *twice differentiable*. If U is twice differentiable, standard uniqueness and existence theorems tell us that there is one and only one curve through each point in R^{2N} satisfying these equations. The vector field $(A_1^0, \ldots, A_N^0, A_1, \ldots, A_N)$ is called the *infinitesimal generator* of U. Hence there is a one–one correspondence between twice differentiable one-parameter groups and differentiable vector fields. Since we already know that $v_i = dq_i/dt$ these equations reduce to

$$dq_i/dt = v_i, \qquad dv_i/dt = A_i(t, q_1, \ldots, q_N, v_1, \ldots, v_N), \qquad i = 1, \ldots, N.$$

This is equivalent to the second order differential equations

$$\frac{d^2 q_i}{dt^2} = A_i\left(t, q_1, \ldots, q_N, \frac{dq_1}{dt}, \ldots, \frac{dq_N}{dt}\right), \qquad i = 1, \ldots, N.$$

We now make further assumptions about our physical laws to restrict the nature of the functions A_i.

1. The A_i's are functions of the q_k's alone and are independent of t and the v_k's.
2. There are positive constants M_i and a function $V(q_1, \ldots, q_N)$ such that $A_i = -(M_i)^{-1}(\partial V/\partial q_i)$, $i = 1, \ldots, N$.

The first requirement assumes that the forces are time and velocity independent while the second follows from assuming there are no dissipative forces, such as friction. M_i is called the *mass* associated with the *i*th coordinate, $p_i = M_i v_i$ the *momentum* component conjugate to q_i, $M_i A_i$ the *force* in the *i*th direction, and V the *potential energy* of the system. Systems satisfying Assumption 2 are called *conservative*. Since the v_i and p_i determine each other uniquely we may regard the state of our system as described by the q_i's and p_i's instead of the q_i, and v_i's. Of course, when this is done S becomes a different subset of $2N$-dimensional space which we call *phase space*. Following tradition we shall now assume that S is phase space. A *dynamical variable* is a function on S. These correspond to quantities which can be measured. An *integral* of our system is a differentiable dynamical variable which is constant on orbits of the dynamical group U.

Lemma 3.2. *The function ϕ is an integral if and only if*

$$\sum_1^N (p_i/M_i) \frac{\partial \phi}{\partial q_i} = \sum_1^N \frac{\partial V}{\partial q_i} \frac{\partial \phi}{\partial p_i}.$$

Proof. The function ϕ is an integral if and only if $(d/dt)\phi(U_t(s)) = 0$, $t \in (-\infty, \infty)$. This last condition is true if and only if

$$0 = \frac{d}{dt} \phi(q_1(t), \ldots, q_N(t), p_1(t), \ldots, p_N(t))$$

$$= \sum \frac{\partial \phi}{\partial q_i} \frac{dq_i}{dt} + \sum \frac{\partial \phi}{\partial p_i} \frac{dp_i}{dt} = \sum (p_i/M_i) \frac{\partial \phi}{\partial q_i} - \sum \frac{\partial V}{\partial q_i} \frac{\partial \phi}{\partial p_i}.$$

More generally, let W be any twice differentiable one-parameter group with infinitesimal generator $(B_1{}^W, \ldots, B_N{}^W, C_1{}^W, \ldots, C_N{}^W)$ on S. Then ϕ is constant on W orbits if and only if

$$\sum B_i{}^W \frac{\partial \phi}{\partial q_i} + \sum C_i{}^W \frac{\partial \phi}{\partial p_i} = 0.$$

If ϕ satisfies:

$$-\frac{\partial \phi}{\partial q_i} = C_i{}^W, \qquad \frac{\partial \phi}{\partial p_i} = B_i{}^W, \qquad i = 1, \ldots, N. \tag{3.2}$$

then ϕ is constant on W orbits. If there is a ϕ satisfying (3.2) then W is called a *one-parameter group of contact transformations* and its infinitesimal generator is called an *infinitesimal contact transformation*. The function ϕ determines W uniquely and is uniquely determined by W up to an additive constant. We call ϕ the *fundamental invariant* of W.

We now show that our dynamical group U is a one-parameter group of contact transformations and hence has at least one nontrivial integral. We must find a function H such that

$$\frac{\partial H}{\partial q_i} = \frac{\partial V}{\partial q_i}, \qquad \frac{\partial H}{\partial p_i} = p_i/M_i, \qquad i = 1, \ldots, N. \qquad (3.3)$$

From the second half of Eqs. (3.3) we have

$$H = \sum p_i^2/2M_i + H_0(q_1, \ldots, q_N),$$

and from the first half $H_0 = V$. Thus $H = \sum p_i^2/2M_i + V(q_1, \ldots, q_N)$ is constant on U orbits and is the fundamental invariant of the one-parameter group of contact transformations U. H is called the *energy* of the system. The fact that H is an integral gives us the law of conservation of energy. $\sum p_i^2/2M_i$ is the *kinetic energy* of the system. H is also called the *Hamiltonian* of the system and we may now write our differential equations in *Hamiltonian form*:

$$\frac{dq_i}{dt} = \frac{\partial H}{\partial p_i}; \qquad \frac{dp_i}{dt} = -\frac{\partial H}{\partial q_i}.$$

Let W be a one-parameter group of contact transformations whose fundamental invariant is ψ, and suppose ϕ is a constant on W orbits. Then

$$\sum \frac{\partial \phi}{\partial q_i} \frac{\partial \psi}{\partial p_i} - \sum \frac{\partial \phi}{\partial p_i} \frac{\partial \psi}{\partial q_i} = 0.$$

The left side of the above equation is called the *Poisson bracket* of ϕ and ψ and is denoted by $[\phi, \psi]$. Notice $[\phi, \psi] = -[\psi, \phi]$; hence ϕ is a constant on the orbits of the one-parameter group of contact transformations with fundamental invariant ψ if and only if ψ is a constant on the orbits of the one-parameter group of contact transformations with fundamental invariant ϕ. Let H be the fundamental invariant of the dynamical group U. A one-parameter group of contact transformations W is called a *one-parameter group of symmetries* if H is constant on W orbits. Let ψ be the fundamental invariant of W. Then ψ is an integral of the dynamical system if and only if W is a one-parameter group of symmetries.

A *momentum integral* is an integral which is a linear function of the momenta, that is, of the form $\sum p_i D_i(q_1, \ldots, q_N)$. It follows from Lemma 3.2 that if V is independent of q_j, then p_j is a momentum integral. This gives us the law of conservation of momentum. Nontrivial momentum integrals do not always exist, but when they do they are important. Let S_1 be the subset of R^N consisting of points $(q_1(t), \ldots, q_N(t))$, $t \in (-\infty, \infty)$. S_1 is called *configuration space*. Let W_t be a twice-differentiable one parameter group in S_1 with infinitesimal generator $(D_1, \ldots, D_N)$. Let $\hat{W}_t$ be the one-parameter group in S whose infinitesimal generator is

$$\left(D_1, \ldots, D_N, -\sum_1^N p_i \frac{\partial D_i}{\partial q_1}, \ldots, -\sum_1^N p_i \frac{\partial D_i}{\partial q_N} \right).$$

Then $\hat{W}_t$ is a one-parameter group of contact transformations whose fundamental invariant is $\sum_1^N p_i D_i(q_1, \ldots, q_N)$. If the latter is an integral then W_t has led us to a momentum integral. For example, suppose we have two interacting particles. Then $s \in S$ has the form

$$s = (x_1, \ldots, x_6, p_1, \ldots, p_6)$$

and

$$H = \sum_1^6 p_i^2/2M_i + V(x_1, \ldots, x_6).$$

Suppose V depends only on the distance between the particles. That is,

$$V(x_1, \ldots, x_6) = V_0([(x_1 - x_4)^2 + (x_2 - x_5)^2 + (x_3 - x_6)^2]^{1/2}) = V_0(r).$$

Then ϕ is an integral if and only if

$$\sum_1^6 \frac{\partial \phi}{\partial x_i} p_i/M_i - r^{-1/2} \frac{\partial V_0}{\partial r} \left[\sum_1^6 \left(\frac{\partial \phi}{\partial p_i} - \frac{\partial \phi}{\partial p_{i+3}} \right)(x_i - x_{i+3}) \right] = 0. \quad (3.4)$$

To find the momentum integrals we look for one-parameter groups in $(x_1, \ldots, x_6)$. Let us first consider a translation in the first coordinate direction represented by

$$A_t(x_1, \ldots, x_6) = (x_1 + t, x_2, x_3, x_4 + t, x_5, x_6).$$

It is easy to check that A_t is a twice differentiable one-parameter group with infinitesimal generator

$$\left. \frac{d}{dt} A_t \right|_{t=0} = (1, 0, 0, 1, 0, 0).$$

We form the one-parameter group of contact transformations $\hat{A}_t$ with infinitesimal generator $(1, 0, 0, 1, 0, \ldots, 0)$ and fundamental invariant

$p_1 + p_4$. Then $\phi = p_1 + p_4$ satisfies (3.4) and is thus a momentum integral. In the same way $p_2 + p_5$ and $p_3 + p_6$ are momentum integrals. We thus have conservation of momentum in the three coordinate directions.

Let us now consider a rotation about the x_3 axis, represented by:

$$B_t(x_1, \ldots, x_6) = (x_1 \cos t + x_2 \sin t, -x_1 \sin t + x_2 \cos t, x_3, x_4 \cos t$$
$$+ x_5 \sin t, -x_4 \sin t + x_5 \cos t, x_6).$$

Again B_t is a twice differentiable one-parameter group with infinitesimal generator

$$\left. \frac{dB_t}{dt} \right|_{t=0} = (x_2, -x_1, 0, x_5, -x_4, 0).$$

We form the one-parameter group of contact transformations B_t with infinitesimal generator

$$(x_2, -x_1, 0, x_5, -x_4, 0, p_2, -p_1, 0, p_5, -p_4, 0),$$

and fundamental invariant

$$p_1 x_2 - p_2 x_1 + p_4 x_5 - p_5 x_4.$$

We again check that this satisfies (3.4), and is thus a momentum integral which gives us conservation of angular momentum.

This method can be generalized to the following. Let $t \to A_t$ be a one-parameter group of rigid motions in R^3 (i.e., transformations which leave distances fixed). Define

$$B_t(x_1, \ldots, x_6) = (A_t(x_1, x_2, x_3), A_t(x_4, x_5, x_6)).$$

Then B_t is a one-parameter group of contact transformations and its fundamental invariant is a momentum integral. Notice that momentum integrals result from a "motion" in the coordinate direction. This will be important when we define momentum integrals in quantum mechanics.

IV. The Quantum Mechanical Logic

A. Propositions, States, and Observables

Let us begin by considering the classical postulates for quantum mechanics due mainly to von Neumann [59]. These are the postulates usually used by physicists, and they form a basis for the more refined and sophisticated postulates of quantum field theory [26, 54]. Unlike the axiomatic systems mentioned in the Introduction, classical quantum mechanics takes the states as its axiomatic elements. The three main axioms are:

(C-1) *The pure states of a quantum mechanical system are vectors in a Hilbert space H.*

(C-2) *The quantum mechanical observables are represented by self-adjoint operators on H.*

(C-3) *The probability that an observable A has a value in a Borel set E when the system is in the state ϕ is $\langle \phi, P^A(E)\phi \rangle$ where $P^A(\cdot)$ is the resolution of identity for A.*

These axioms are unfortunately *ad hoc* and have little physical significance. They also contain other deficiencies mentioned in the Introduction. For these reasons we shall attempt to replace these axioms by axioms with physical significance. We shall formulate the axioms in such a way that at least a generalized form of modern probability theory is applicable. Furthermore we shall show that our axioms are "close" to the classical ones (C-1), (C-2), (C-3), and in a certain weak statistical sense reduce to these classical axioms.

Since our goal is to establish a probabilistic theory, let us first consider a familiar probabilistic system. Suppose we have a classical mechanical system consisting of a large number n of particles. We would describe this system by points $(q_1, \ldots, q_{3n}, p_1, \ldots, p_{3n})$ in phase space S which we called states. However, if n is very large (about 10^{23} molecules in a liter of air) it would be impossible to determine the exact state of our system. The best we can do is make statistical predictions about the state using the macroscopic knowledge we have about the system such as temperature, pressure, volume, etc. Now, instead of the state being given by a point in S it is described by a probability measure on S which gives the probability that the system is in a given region of S. Of course, the more precise the data concerning our system is the more concentrated our probability measure becomes until ultimately the measure is concentrated at a single point and we are back to our point description of states.

Now suppose we have a quantum mechanical particle p such as an electron. Proceeding classically we form a phase space $S = (x, y, z, p_x, p_y, p_z)$. However a point in S now has no physical meaning since the Heisenberg uncertainty principle informs us that there is no way of telling if a particle is at the point $(x_0, y_0, z_0, p_{x_0}, p_{y_0}, p_{z_0})$. Therefore the fundamental principle (C) of classical mechanics does not apply. If we know the state of p approximately at time t_0, that is we know p is in a certain region of S, we can only make statistical predictions about where it will be at time $t_1 > t_0$. Thus instead of the points of S being important as they are in classical mechanics it is the subsets of S which become important in quantum physics while the points are ignored altogether. Also one must be careful not to consider sets in S which are too small and thus violate the uncertainty principle. These allowed subsets will be called propositions (they are also referred to as events and questions in the literature) and our axiomatic model will be based upon them.

Having considered the above motivating preliminaries, let us proceed to the development of an axiomatic model. Let $L = \{a, b, c, \ldots\}$ be the set of *propositions* (*events, questions*) of a quantum mechanical system. A proposition may be thought of physically as a statement concerning the system for which there is an experiment that can verify whether the statement is true or false. For example, " The electron has x coordinate between 2 and 3 cm and x momentum component between 1 and 2 g cm sec^{-1}." might be a proposition. We now give axioms which the set L should satisfy. Whenever a proposition a is true it follows that a proposition b is also true and we say a *implies* b and write $a \leq b$. This relation should satisfy

(L-1) $a \leq a$, for all $a \in L$,
(L-2)   if $a \leq b$, $b \leq c$, then $a \leq c$,
(L-3)   if $a \leq b$, $b \leq a$, then $a = b$.

[The equal sign in (L-3) indicates that a and b are experimentally indistinguishable. Thus L becomes a partially ordered set under $\leq$. We assume there is an *absurd proposition* 0 which is never true, (e.g., the electron is nowhere) and a self-evident proposition 1 which is always true (e.g., the electron is somewhere).]

(L-4)   There are propositions 0, 1 satisfying $0 \leq a \leq 1$ for all $a \in L$.

In any partially ordered set we can define suprema and infima. The *supremum* (or *least upper bound*) of a and b written $a \vee b$, if it exists, is the unique proposition which satisfies $a \vee b \geq a, b$ and if $c \geq a, b$, then $c \geq a \vee b$. Similarly the *infimum* (or *greatest lower bound*) of a and b written $a \wedge b$, if it exists, is the unique proposition which satisfies $a \wedge b \leq a, b$ and if $c \leq a, b$ then $c \leq a \wedge b$. Now $a \vee b$ may be interpreted as the proposition which is true when a or b is true and $a \wedge b$ as the proposition which is true when a and b are true. Note that we do not assume that the supremum or infimum of two elements in L necessarily exist.

Given any $a \in L$ we assume that there is an $a' \in L$ which is true whenever a is not true. We call a' the *complement* of a and postulate that the map $a \to a'$ satisfies:

(L-5) $(a')' = a$ for all $a \in L$,
(L-6) if $a \le b$, then $b' \le a'$,
(L-7) $a \vee a' = 1$ for all $a \in L$.

The reader should note that these axioms are intuitively evident. It is easily checked that if $a \vee b$ exists then $a' \wedge b'$ exists and $(a \vee b)' = a' \wedge b'$. Similarly if $a \wedge b$ exists, then $a' \vee b'$ exists and $(a \wedge b)' = a' \vee b'$. These are de Morgan's laws. If $a \le b'$ we say that a and b are *disjoint* and write $a \perp b$. Notice that this is a symmetrical relation; that is $a \perp b$ if and only if $b \perp a$. It is trivial that $a \perp b$ implies $a \wedge b = 0$, although the converse is not true. It is also trivial that if $\{a_\lambda\}$ and $\{b_\mu\}$ are two indexed sets of propositions for which $a_\lambda \perp b_\mu$ for all λ and μ and $\vee a_\lambda$ and $\vee b_\mu$ exist, then $\vee a_\lambda \perp \vee b_\mu$. If $a \perp b$, for convenience we write $a + b$ for $a \vee b$, if the supremum exists. We may interpret $a \perp b$ to mean a is true implies b is false, so it is physically reasonable to assume that $a + b$ exists. We now extend this to a countable number of disjoint propositions.

(L-8) If a_i is a sequence of mutually disjoint propositions then $\sum a_i$ exists.

We now make this into a statistical system by introducing states. A state should give us the condition of our system and should describe our system as completely as possible. That is, given a proposition a, the state should tell us the probability that a is true. We thus define a *state* m as a map from L into the unit interval $[0, 1] \subset R$ which satisfies:

(M-1) $m(1) = 1$.
(M-2) $m(\sum_1^\infty a_i) = \sum_1^\infty m(a_i)$.

Notice that if m_i is a sequence of states and $\lambda_i \ge 0$, $\sum_1^\infty \lambda_i = 1$, then $m = \sum_1^\infty \lambda_i m_i$ is a state where $m(a) = \sum \lambda_i m_i(a)$ for all $a \in L$. The state m is called a *mixture* of the m_i's, and m is interpreted physically as being the state in which we only know that the system is in state m_i

with probability λ_i. A state which is not a mixture is *pure*. A set of states M on L is *full* if M is closed under mixtures and

(F-1) If $a \neq 0$ there is an $m \in M$ such that $m(a) = 1$;
(F-2) $m(a) = m(b)$ for every $m \in M$ implies $a = b$.

One can give examples of systems which have no states at all [3, p. 186]. For this reason we postulate:

(L-9) There is a full set of states M on L.

If the pair (L, M) satisfies (L-1)–(L-9) it is called a *logic*. We sometimes abuse this notation by calling L a logic and M a full set of states on L. We say that two propositions a, b are *compatible* (written $a \leftrightarrow b$) if there are mutually disjoint propositions a_1, b_1, c such that $a = a_1 + c$, $b = b_1 + c$. Compatible propositions are ones which physically may be verified at the same time. Another way of saying this is that compatible propositions are ones whose corresponding experiments do not inter- fere with each other. We can briefly see this as follows. First of all dis- joint propositions are one of the simplest examples of noninterfering propositions. This is because if one is true the other is false and one experiment may be used to verify them simultaneously. Thus if a_1, b_1, c are mutually disjoint they correspond to noninterfering experiments and the same would be true of $a = a_1 + c$ and $b = b_1 + c$. Conversely, suppose a,b correspond to noninterfering experiments. Then $c = a \wedge b$ certainly should exist and so should $a_1 = a \wedge [(a \wedge b)']$ and $b_1 = b \wedge [(a \wedge b)']$. But now one can easily convince oneself that a_1, b_1, and c are mutually disjoint and $a = a_1 + c$, $b = b_1 + c$.

Notice that if $a \perp b$, then $a \leftrightarrow b$ since $a = a + 0$, $b = b + 0$. Physically one would expect if $a \leq b$ then $a \leftrightarrow b$. Our first lemma proves this.

Lemma 4.1. *Let* (L, M) *be a logic and suppose* $a \leq b$ *in* L. *Then there is a unique element* c *such that* $a + c = b$ *and hence* $a \leftrightarrow b$.

Proof. Since $a \perp b'$, $a + b'$ exists and hence using de Morgan's law $a' \wedge b = (a + b')'$ exists. Now $a \perp (a' \wedge b)$ and hence $a + (a' \wedge b)$ exists. Now $m(a + a' \wedge b) = m(a) + m(a' \wedge b) = m(a) + 1 - m(a + b')$ $= m(b)$ and since this holds for every $m \in M$, $b = a + (a' \wedge b)$. Suppose $b = a + c$. Since $m(c) = m(b) - m(a) = m(a' \wedge b)$ we have $c = a' \wedge b$ and thus c is unique. The last statement holds since $b = c + a$ and $a = 0 + a$.

We denote this unique element $c = b \wedge a'$ by $b - a$.

Lemma 4.2. *Let (L, M) be a logic, $a,b \in L$ and suppose $a \leftrightarrow b$. Then $a \leftrightarrow b'$, and $a \vee b$, $a \wedge b$ exist. If a_1, b_1, c are mutually disjoint with $a = a_1 + c$, $b = b_1 + c$, then a_1, b_1, c are unique. In fact $c = a \wedge b$, and*

$$a_1 = a - a \wedge b = a \vee b - b, \qquad b_1 = b - a \wedge b = a \vee b - a.$$

Proof. Let $d = a_1 + b_1 + c$. Now $a \leq d$ and $b \leq d$. If $a \leq e$ and $b \leq e$, then $d = a_1 + b_1 + c \leq e$. Thus $a \vee b$ exists and $a \vee b = d$. Now $d = a + b_1 = b + a_1$ and so $a_1 = d - b = a \vee b - b$ and $b_1 = d - a = a \vee b - a$. Moreover, $a = d - b_1$ and $b = d - a_1$ so $a' = b_1 + d'$ and $b' = a_1 + d'$. Thus $a \leftrightarrow b'$ and by the above argument $a' \vee b' = a_1 + b_1 + d' = [d - (a_1 + b_1)]' = c'$. Therefore $c = a \wedge b$.

Corollary 4.3. $a \leftrightarrow b$ *if and only if* $a \wedge b$ *exists and* $(a - a \wedge b) \perp b$.

Proof. If $a \leftrightarrow b$ then $a \wedge b$ exists and $(b - a \wedge b) \leq (a - a \wedge b)'$ by Lemma 4.2. Since $a \wedge b \leq (a - a \wedge b)'$ we have $b = (b - a \wedge b) + (a \wedge b) \leq (a - a \wedge b)'$. Hence $(a - a \wedge b) \perp b$. Conversely if $a \wedge b$ exists and $(a - a \wedge b) \perp b$ then $(b - a \wedge b) \perp (a - a \wedge b)$. Hence $a = (a - a \wedge b) + a \wedge b$ and $b = (b - a \wedge b) + a \wedge b$ and thus $a \leftrightarrow b$.

Dynamical variables are very important in classical systems and they are equally important in quantum mechanics. Since we have no phase space, however, they must be defined differently. To distinguish these objects from their classical counterpart we shall call them *observables*. In order to define observables, let us consider the properties they should possess. An observable should be an object associated with our system which can be measured. That is, it determines a set of real numbers, the values of the observable. On the other hand, given any set of real numbers E an observable x gives us the proposition: x has a value in E. We thus define an observable x as a map from the Borel subsets $\mathcal{B}(R)$ of the real line R into L which satisfies:

(O-1) $x(R) = 1$
(O-2) If $E \cap F = \emptyset$, then $x(E) \perp x(F)$
(O-3) $x(\cup E_i) = \sum_1^\infty x(E_i)$ if $E_i \cap E_j = \emptyset$, $i \neq j$.

It follows that $x(\emptyset) = 0$ and denoting the complement of $E \in \mathcal{B}(R)$ by E' we have $x(E') = x(E)'$. To give an example of an observable, let a_i be a sequence of mutually disjoint propositions such that $\sum a_i = 1$ and let λ_i be a sequence of distinct real numbers. Defining the map x by $x(E) = \sum \{a_j : \lambda_j \in E\}$, $E \in \mathcal{B}(R)$, it is easily checked that x is an observable. Two observables x,y are *compatible* (written $x \leftrightarrow y$) if $x(E) \leftrightarrow y(F)$ for every $E,F \in \mathcal{B}(R)$. Observables which are compatible may be thought of physically as being observables which are simultaneous measurable. We shall show in the next section that a collection of compatible observables may be identified with a collection of random variables.

The reader should notice that we have constructed a generalized probability theory. Instead of being a σ-algebra of subsets of a set our events, which are more general, form a logic with less structure than a σ-algebra. The probability measures are replaced by states and the random variables by observables. Before proceeding further, let us consider two examples of logics.

Example 1. Let (Ω, F) be a measurable space. The σ-algebra F satisfies (L-1)–(L-8) and from the existence of measures concentrated at points we see that F has a full set of states and is thus a logic. If x is an observable it follows from a theorem of Sikorski [52] that there exists an F-measurable function f such that $x(E) = f^{-1}(E)$ for all $E \in \mathscr{B}(R)$. Thus observables are just inverses of measurable functions. This example gives the conventional Kolmogorov formulation of probability theory. It is easily checked that all events and observables are compatible in this example.

Example 2. Let H be a separable Hilbert space and let P be the collection of all closed subspaces of H. Ordering P by inclusion and defining the complement of a subspace as its orthocomplement it is easily seen that P satisfies (L-1)–(L-8). If $a \in P$ we denote the unique orthogonal projection on a by p_a. Now if $\phi \in H$ and $\|\phi\| = 1$ then the map $a \to \langle \phi, p_a \phi \rangle$ is a state. If $a \neq b$ choose a unit vector ϕ_0 in a which is not in b. Then $\langle \phi_0, p_a \phi_0 \rangle = 1$ and $\langle \phi_0, p_b \phi_0 \rangle \neq 1$ so L has a full set of states and is thus a logic. It is an interesting and important fact that every state is a convex combination of states of the above type. Indeed Gleason [12] has shown that any state m on P has the form $m(a) = \sum \lambda_i \langle \phi_i, p_a \phi_i \rangle$ where ϕ_i is an orthonormal set of vectors. Identifying closed subspaces with their orthogonal projections, an observable may be thought of as a projection-valued measure. Since using the spectral theorem, there is a one–one correspondence between projection-valued measures and self-adjoint operators we may identify observables with self-adjoint operators. We shall show later that two propositions are compatible if and only if the corresponding projections commute and two observables are compatible if and only if the corresponding self-adjoint operators commute. This example gives us the usual framework of classical quantum mechanics resulting in the three axioms (C-1), (C-2), and (C-3) stated at the beginning of this subsection.

Lemma 4.4. *Two propositions a, b are compatible if and only if they are in the range of a single observable.*

Proof. If a and b are compatible there are mutually disjoint propositions a_1, b_1, c such that $a = a_1 + c$, $b = b_1 + c$. Define the observable x by $x(\{0\}) = a_1$, $x(\{1\}) = b_1$, $x(\{2\}) = c$, and $x(\{3\}) = (a_1 + b_1 + c)'$. Then $a = x(\{0, 2\})$, $b = x(\{1, 2\})$ and thus a and b are in the range of x. Conversely, suppose there is an observable x and Borel sets E, F such that $a = x(E)$, $b = x(F)$. Then $x(E \cap F)$, $x(E \cap F')$, and $x(F \cap E')$ are mutually disjoint propositions and $a = x(E) = x(E \cap F') + x(E \cap F)$, $b = x(F) = x(F \cap E') + x(E \cap F)$ so $a \leftrightarrow b$.

This last lemma reinforces our definition of compatibility since to verify two compatible propositions we need measure only a single observable. If x is an observable and u a Borel function on R then there is an operational significance for $u(x)$. That is, if x has the value $\lambda \in R$ then $u(x)$ has the value $u(\lambda)$. This is equivalent to saying that the proposition "x has a value in $u^{-1}(E)$" is the same as the proposition "$u(x)$ has a value in E." Motivated by this we define $u(x)$ as $u(x)(E) = x(u^{-1}(E))$ for all $E \in \mathscr{B}(R)$. It is easily checked that $u(x)$ is an observable and that $u(x) \leftrightarrow x$.

B. Boolean Sub σ-Algebras and Compatibility

In the sequel (L, M) will denote a logic. Most of the results in this subsection are due to Varadarajan [57] and Pool [41].

Lemma 4.5. *Let $\{a, a_1, a_2, \ldots\} \subset L$, suppose $a \leftrightarrow a_i$, $i = 1, 2, \ldots$, and suppose $\vee a_i$ and $\vee (a \wedge a_i)$ exist. Then (i) $a \leftrightarrow \vee a_i$ and (ii) $a \wedge (\vee a_i) = \vee (a \wedge a_i)$.*

Proof. Let $c = \vee (a \wedge a_i)$. Clearly $c \leq a$ and $c \leq \vee a_i$. Since $a \leftrightarrow a_i$, it follows from Lemma 4.2 that $a - (a \wedge a_i) \leq a_i'$. Since $a \wedge a_i \leq c$ we have $a - c \leq a - (a \wedge a_i) \leq a_i'$. Hence $(a - c) \perp a_i$ and $(a - c) \perp \vee a_i$. Therefore $(a - c) \perp (\vee a_i - c)$, and we have $a = c + (a - c)$ and $\vee a_i = c + (\vee a_i - c)$. It follows that $a \leftrightarrow \vee a_i$ and from Lemma 4.2 we have $c = \vee (a \wedge a_i) = a \wedge (\vee a_i)$.

We call L a *lattice* if the infimum and supremum of any two elements in L exist, and L is *distributive* if it is a lattice and the *distributive law* $a \wedge (b \vee c) = (a \wedge b) \vee (a \wedge c)$ holds for all $a, b, c \in L$. A distributive logic L is a Boolean σ-algebra. A subset B of L is a *Boolean sub σ-algebra* if B is a distributive logic when the infimum and supremum operations on L are restricted to B. Since $\mathscr{B}(R)$ is a Boolean σ-algebra it follows that the range of any observable is a Boolean sub σ-algebra.

Theorem 4.6. *A logic L is a Boolean σ-algebra if and only if its elements are mutually compatible.*

Proof. If L is a Boolean σ-algebra then $a - a \wedge b = a \wedge (a \wedge b)' = a \wedge (a' \vee b') = a \wedge b' \leq b'$. Since $(a - a \wedge b) \perp b$ it follows from Lemma 4.3 that $a \leftrightarrow b$. Conversely if all the elements of L are mutually compatible, then by Lemma 4.2 L is a lattice and by Lemma 4.5 L satisfies the distributive law.

We thus see that Boolean σ-algebras give logics in which no two experiments interfere. Since interfering experiments are characteristic of quantum mechanical effects one would expect that Boolean σ-algebras describe purely classical phenomena and, in fact, this is true. Applying a representation theorem of Loomis [28], if B is a Boolean σ-algebra there is a measurable space (Ω, F) and a map h from F onto B satisfying

(i) $h(\Omega) = 1$;
(ii) if $\Lambda \cap \Gamma = \varnothing$, then $h(\Lambda) \perp h(\Gamma)$;
(iii) $h(\cup \Lambda_i) = \sum h(\Lambda_i)$ if $\Lambda_i \cap \Lambda_j = \varnothing, i \neq j$.

The map h is called a σ-*homomorphism*. Notice that this is a generalization of an observable which is a σ-homomorphism based on $\mathscr{B}(R)$. Now the measurable space (Ω, F) may be thought of as a phase space for some classical mechanical system and thus we are back to classical physics. If m is a state on B then m induces a probability measure $\mu: \Lambda \rightarrow m(h(\Lambda))$, $\Lambda \in F$ and thus we have a probability space (Ω, F, μ). It is of value to find out what the observables reduce to in this case. If x is an observable with range in B, then it follows from Sikorski's theorem that there exists an F measurable function f such that $x(E) = h[f^{-1}(E)]$. Thus observables reduce to random variables, and our generalized probability theory reduces to the conventional Kolmogorov probability theory on a classical phase space.

If $S \subset L$ the *compatant* of S is $S^c = \{b \in L: b \leftrightarrow a \text{ for all } a \in S\}$. Clearly $S_1 \subset S_2$ implies $S_2{}^c \subset S_1{}^c$. Denoting $(S^c)^c$ by S^{cc} it is clear that

$S \subset S^{cc}$. A set $S \subset L$ is *compatible* if the elements of S are mutually compatible. Clearly $S \subset S^c$ if and only if S is compatible. If $A_1, A_2 \subset L$ we write $A_1 \leftrightarrow A_2$ if $a_1 \leftrightarrow a_2$ for all $a_1 \in A_1$, $a_2 \in A_2$. It is important to know when a collection of Boolean sub σ-algebras is contained in a single Boolean sub σ-algebra.

Theorem 4.7. *If B_1 and B_2 are Boolean sub σ-algebras, there is a Boolean sub σ-algebra containing them if and only if $B_1 \leftrightarrow B_2$.*

Proof. We first assume that B_1 and B_2 have a finite number of elements and $B_1 \leftrightarrow B_2$. Let $\{s_i \colon i = 1, \ldots, p\}$ and $\{t_i \colon i = 1, \ldots, q\}$ be the distinct nonzero minimal elements of B_1 and B_2, respectively. The s_i's are mutually disjoint since otherwise $s_i \wedge s_j \neq 0$ for some i, j and $s_i \wedge s_j = s_i = s_j$ a contradiction. Also $\bigvee s_i = 1$ since otherwise we would be missing a minimal element. We also have the same properties for the t_i's. Let $c_{ij} = s_i \wedge t_j$, $i = 1, \ldots, p, j = 1, \ldots, q$. If $c_{ij} \neq c_{i'j'}$ then either $i \neq i'$ or $j \neq j'$ and hence either $c_{ij} \leq s_i$ and $c_{i'j'} \leq s_{i'}$, $i \neq i'$, or $c_{ij} \leq t_j$ and $c_{i'j'} \leq t_{j'}$, $j \neq j'$. Hence the c_{ij}'s are disjoint. Also $\sum_{j=1}^{q} c_{ij} = \bigvee \{s_i \wedge t_j \colon j = 1, \ldots, q\} = s_i \wedge (\vee t_j) = s_i$. Similarly $\sum_{i=1}^{p} c_{ij} = t_j$ and $\sum_{i,j=1}^{p,q} c_{ij} = 1$. If B is 0 together with all suprema of the c_{ij}'s, then $B_1 \subset B$ and $B_2 \subset B$, and it is trivial that B is a Boolean sub σ-algebra with a finite number of elements. Denote B by $[B_1, B_2]$.

Now assume that B_1 and B_2 are arbitrary Boolean sub σ-algebras and $B_1 \leftrightarrow B_2$. Let $B = \bigcup [A_1, A_2]$ where A_1 and A_2 run over all finite Boolean sub σ-algebras such that $A_1 \subset B_1$ and $A_2 \subset B_2$. Clearly $B_1 \subset B$ and $B_2 \subset B$. Now $0, 1 \in B$ and if $a \in B$, then $a' \in B$. If $a, b \in B$, then there are finite Boolean sub σ-algebras A_1, A_2, A_1^0, A_2^0 such that $a \in [A_1, A_2]$, $b \in [A_1^0, A_2^0]$. Letting $A_1^{00} = [A_1, A_1^0]$ and $A_2^{00} = [A_2, A_2^0]$ we see that a and b are in $[A_1^{00}, A_2^{00}]$. Thus $a \vee b \in B$ and $a \leftrightarrow b$. It follows that B is a Boolean subalgebra (closed under *finite* suprema and infima). Now by Zorn's lemma, B is contained in a maximal Boolean subalgebra B_0. Let a_i be a sequence of mutually disjoint elements of B_0. Let $a = \sum a_i$ and let $b \in B_0$. Then $b \leftrightarrow a_i$, $i = 1, 2, \ldots$, and hence by Lemma 4.5 $b \leftrightarrow a$. Letting $C = \{0, 1, a, a'\}$, since B_0 is maximal we have $B_0 = [B_0, C]$ and thus $a \in B_0$. Thus B_0 is a Boolean sub σ-algebra and the proof is complete.

The following example, due to Ramsey [43], shows the surprising fact that Theorem 4.7 does not hold for more than two Boolean sub σ-algebras. If $L = \{a \colon a$ is a subset of $\{1, 2, \ldots, 8\}$ with an even number of elements$\}$, then L is a logic under the usual operations. Let $a = \{1, 2, 3, 4\}$, $b = \{1, 3, 5, 6\}$, and $c = \{1, 3, 6, 8\}$. Then a, b, and c

are mutually compatible but $(a \vee b) \wedge c$ does not exist, since $\{1, 3\}$ and $\{1, 6\}$ are lower bounds of $\{a \vee b, c\}$ and there is no lower bound larger than both of them. Thus $a \vee b \leftrightarrow c$ and $\{a, b, c\}$ is not contained in a Boolean sub σ-algebra.

However, there is a simple, necessary and sufficient condition for Theorem 4.7 to hold for any collection of Boolean sub σ-algebras. L satifies *condition* C (for compatibility) if for any three mutually compatible elements a, b, c we have $a \leftrightarrow b \vee c$. It follows from Lemma 4.5 that if L is a lattice it satisfies condition C.

Theorem 4.8. *Any compatible set $S \subset L$ is contained in a Boolean sub σ-algebra if and only if L satisfies condition C.*

Proof. Necessity is trivial so we turn to sufficiency, and assume L satisfies condition C. Since $S \subset S^{cc}$ it suffices to show that S^{cc} is a Boolean sub σ-algebra. Obviously $0, 1 \in S^{cc}$ and if $a \in S^{cc}$ then $a' \in S^{cc}$. Now suppose a_i is a sequence of mutually disjoint elements of S^{cc}, and suppose $b \in S^c$. Then $a_i \leftrightarrow b$, $i = 1, 2, \ldots$ so by Lemma 4.5 $\sum a_i \leftrightarrow b$ and thus $\sum a_i \in S^{cc}$. If b_1 and b_2 are in S^{cc} then by condition C, $b_1 \vee b_2 \in S^{cc}$ so S^{cc} is a lattice. Now from $S \subset S^c$ we get $S^{cc} \subset S^c$ and hence $S^{cc} \subset (S^{cc})^c$. Thus S^{cc} is compatible and again by Lemma 4.5 the distributive law holds so S^{cc} is a Boolean sub σ-algebra.

Corollary 4.9. *Any collection $\{B_\lambda\}$ of mutually compatible Boolean sub σ-algebras is contained in a Boolean sub σ-algebra if and only if L satisfies condition C.*

Note that there is a subtlety in the proof of Theorem 4.8. One might at first think that after showing S^{cc} is a sublogic it is sufficient to show that S^{cc} is compatible. However, this is not enough. It must be shown that S^{cc} is a lattice and this is where condition C is needed.

A Boolean sub σ-algebra B is *separable* if there is a countable subset C of B such that the smallest Boolean sub σ-algebra containing C is B itself. A logic L is *separable* if every Boolean sub σ-algebra in L is separable. Since $\mathscr{B}(R)$ is separable it easily follows that the range of an observable is a separable Boolean sub σ-algebra. The converse of this statement is proved in [57].

Lemma 4.10. *B is a separable Boolean sub σ-algebra if and only if B is the range of an observable.*

Now let (Ω, F) be a measurable space, f an F-measurable function, and G the smallest σ-algebra with respect to which f is measurable. It is a well-known measure-theoretic fact that if g is measurable with respect to G then g is a Borel function of f. Applying Sikorski's theorem we can now prove:

Lemma 4.11. *If the range of an observable y is contained in the range of an observable x then there is a Borel function u such that $y = u(x)$.*

Denote the range of an observable x by $R(x)$.

Theorem 4.12. (i) *Two observables x_1, x_2 are compatible if and only if there is an observable x and Borel functions u_1, u_2 such that $x_1 = u_1(x)$ and $x_2 = u_2(x)$. (ii) If (L, M) satisfies condition C then a sequence of observables x_i is compatible if and only if there is an observable x and Borel functions u_i such that $x_i = u_i(x)$. (iii) If (L, M) is separable and satisfies condition C, then a collection of observables x_λ is compatible if and only if there is an observable x and Borel functions u_λ such that $x_\lambda = u_\lambda(x)$.*

Proof. (i) Sufficiency is trivial. For necessity suppose $x_1 \leftrightarrow x_2$. Then by Theorem 4.7 there is a Boolean σ-algebra containing $R(x_1) \cup R(x_2)$. Intersecting all the Boolean sub σ-algebras containing $R(x_1) \cup R(x_2)$ we see that there is a smallest Boolean sub σ-algebra B containing $R(x_1) \cup R(x_2)$. Now since $R(x_1)$ and $R(x_2)$ are separable it follows that B is separable and applying Lemma 4.10 there is an observable x such that $R(x) = B$. Applying Lemma 4.11 there are Borel functions u_1 and u_2 such that $x_1 = u_1(x)$, $x_2 = u_2(x)$. The proofs of (ii) and (iii) are similar.

Corollary 4.13. *Under suitable conditions given in Theorem 4.12 a collection (sequence, pair) of observables x_λ is compatible if and only if there is a measurable space (Ω, F), a σ-homomorphism h from F into L and F-measurable functions f_λ such that $x_\lambda(E) = h(f_\lambda^{-1}(E))$ for every $E \in \mathscr{B}(R)$.*

We now consider the important question of joint distributions. By analogy with the conventional definition of joint distributions, we say that two observables x, y have a *joint distribution* if there is a σ-homomorphism z from the Borel sets $\mathscr{B}(R^2)$ of the plane R^2 into L such that $x(E) = z(E \times R)$ and $y(E) = z(R \times E)$ for all $E \in \mathscr{B}(R)$. We call z the *joint homomorphism* of x, y. It is easily seen that z is unique if it exists and that z always exists in the conventional probability theory. It is a remarkable fact that in the generalized theory the joint homomorphism need not exist. If it does exist and $m \in M$ we define the *joint distribution* of x, y under m as $m_{x,y}(E) = m(z(E))$ for all $E \in \mathscr{B}(R^2)$.

Corollary 4.14. *The joint homomorphism of x_1 and x_2 exists if and only if $x_1 \leftrightarrow x_2$.*

Proof. Necessity is trivial. To prove sufficiency suppose $x_1 \leftrightarrow x_2$. By Corollary 4.13 there is a measurable space (Ω, F), a σ-homomorphism $h\colon F \to L$ and F-measurable functions f_1, f_2 such that $x_i(E) = h(f_i^{-1}(E))$, $i = 1$, 2, $E \in \mathscr{B}(R)$. Defining $\phi\colon \Omega \to R^2$ by $\phi(\omega) = (f_1(\omega), f_2(\omega))$, if $\Lambda \in \mathscr{B}(R^2)$ then $\phi^{-1}(\Lambda) \in F$ and $f_1^{-1}(E) = \phi^{-1}(E \times R)$, $f_2^{-1}(E) = \phi^{-1}(R \times E)$ for all $E \in \mathscr{B}(R)$. Now define z by $z(\Lambda) = h(\phi^{-1}(\Lambda))$ for $\Lambda \in \mathscr{B}(R^2)$.

Notice that the joint homomorphism and the above corollary can be extended to an arbitrary number of observables under suitable conditions. In particular if L satisfies condition C then Corollary 4.14 holds for a sequence of observables. We can now use the joint homomorphism to define functions of several compatible observables. Let $x_1, \ldots, x_n$ be compatible observables in a logic satisfying condition C, let z be the joint homomorphism, and let ψ be a Borel function on R^n. We then define $\psi(x_1, \ldots, x_n)$ as the observable $E \to z(\psi^{-1}(E))$. We now show that this definition of a function of several compatible observables has the properties one would expect of a functional calculus.

Corollary 4.15. *If $x_i = u_i(x)$, $i = 1, \ldots, n$ and ψ is a Borel function on R^n, then $\psi(x_1, \ldots, x_n) = \psi(u_1, \ldots, u_n)(x)$.*

Proof. We shall prove the case $n = 2$. There is a measurable space (Ω, F), a σ-homomorphism $h\colon F \to L$ and a F-measurable function f such that $x(E) = h(f^{-1}(E))$. Defining $f_i = u_i \circ f$ we have $x_i(E) = x(u_i^{-1}(E)) = h(f^{-1}u_i^{-1}(E)) = h(f_i^{-1}(E))$, $i = 1$, 2. Defining ϕ and z as in Corollary 4.14 we see that z is the joint homomorphism for x_1, x_2. Hence

$$
\begin{aligned}
\psi(x_1, x_2)(E) &= z(\psi^{-1}(E)) = h(\phi^{-1}(\psi^{-1}(E))) \\
&= h((u_1 \circ f, u_2 \circ f)^{-1}(\psi^{-1}(E))) = h(f^{-1}((u_1, u_2)^{-1}(\psi^{-1}(E))) \\
&= x((u_1, u_2)^{-1}(\psi^{-1}(E))) = x(\psi(u_1, u_2)^{-1}(E)) \\
&= \psi(u_1, u_2)(x)(E).
\end{aligned}
$$

We now give an example of a logic which is closely related to ordinary probability theory yet still retains many quantum mechanical properties. Let Ω be a non-empty set. A σ-class L of subsets of Ω is a collection of subsets which satisfy

(i) $\Omega \in L$; (ii) if $a \in L$, then $a' \in L$; (iii) if a_i are mutually disjoint sets in L, then $\bigcup a_i \in L$, $i = 1, 2, \ldots$.

Following Suppes [54a], if M is the set of states on L, we call (Ω, L, M) a *quantum probability space*. It is easily seen that (L, M) is a logic. If one thinks of Ω as a phase space and L the class of quantum mechanical events concerning coordinate and momenta measurements then one can see intuitively the reasons for (i), (ii), and (iii) in our definition of L. Quantum probability spaces may be used to describe the coordinates and momenta of a quantum mechanical particle but cannot be used for more general systems. In this particular example compatibility has a very simple form. In fact, it is easy to show that $a \leftrightarrow b$ if and only if $a \cap b \in L$.

Observables may be thought of as functions $f \colon \Omega \to R$ such that $f^{-1}(E) \in L$ for all $E \in \mathcal{B}(R)$. If f, g are observables, one might ask whether $f + g$, $f - g$, $f \cdot g$, etc., are observables. It follows from our previous work that if $f \leftrightarrow g$ then $f + g, f - g, f \cdot g$, etc., are observables. Also it is easy to see that L is a σ-algebra if and only if the sum (difference, product) of any two observables is an observable. There are examples of noncompatible observables whose sum is an observable however. Let $\Omega = \{1, 2, 3, 4, 5, 6\}$ and let L be the class of subsets of Ω with an even number of elements. Then L is a σ-class. Define the functions f and g by $f(1) = f(2) = 0$, $f(3) = f(4) = 1$, $f(5) = f(6) = 2$; $g(1) = g(6) = 1$, $g(2) = g(4) = 2$, $g(3) = g(5) = 0$. Then $f \leftrightarrow g$ since $f^{-1}(\{0\}) \cap g^{-1}(\{1\}) = \{1\} \notin L$ for instance. Now it is easy to check that $f + g$ and $f \cdot g$ are observables. However $f - g$ is not an observable!

Let us now briefly consider integration in this theory. If f is an observable and m a state then $A_f = \{f^{-1}(E) \colon E \in \mathcal{B}(R)\}$ is a σ-algebra and m is a measure on A_f. Thus (Ω, A_f, m) becomes an ordinary probability space and we define the integral $\int f \, dm$ in the usual way. It is easy to see that if f and g are any bounded observables such that $\int f \, dm = \int g \, dm$ for every state m, then $f = g$. Another natural question is whether integration is linear. That is, does $\int (f + g) \, dm = \int f \, dm + \int g \, dm$ for f, g bounded observables whose sum is an observable? The author does not know the answer to this question, although it is easily seen to be yes if f and g are simple functions. Other results along these lines may be found in [17b].

C. THE SPECTRUM AND AVERAGE VALUES

The *spectrum* $\sigma(x)$ of an observable x is the smallest closed subset Λ of R such that $x(\Lambda) = 1$. The spectrum of x corresponds physically to the allowable values of x, that is, the values that x may attain. It is easy to show that if L has an infinite number of disjoint nonzero elements, then any nonempty closed subset of R is the spectrum of some observable. It is also clear that $\lambda \in \sigma(x)'$ if and only if there is an open set U containing λ such that $x(U) = 0$. The *point spectrum* $\sigma_p(x)$ of x is $\sigma_p(x) = \{\lambda \in R : x(\lambda) \neq 0\}$, and of course $\sigma_p(x) \subset \sigma(x)$. We call $\sigma_c(x) = \sigma(x) - \sigma_p(x)$ the *continuous spectrum* of x. Suppose λ is an isolated point of $\sigma(x)$. Then there is an $\varepsilon > 0$ such that $E - \{\lambda\} \subset \sigma(x)'$ where $E = \{\omega \in R : |\omega - \lambda| < \varepsilon\}$. Now suppose $\lambda \in \sigma_c(x)$. Then $x(E) = x(\{\lambda\}) + x(E - \{\lambda\}) = 0$ which is impossible. Thus isolated points of $\sigma(x)$ are in $\sigma_p(x)$. If $\sigma(x)$ is a bounded set, we say that x is *bounded* and if x is bounded, we define the *norm* of x to be $|x| = \sup\{|\lambda| : \lambda \in \sigma(x)\}$.

The *distribution* of an observable x in the state m is the probability measures $E \to m[x(E)]$, $E \in \mathscr{B}(R)$. The *expectation* or *average value* of x under m, if it exists, is $m(x) = \int \lambda\, m[x(d\lambda)]$. It follows that if u is a Borel function then $m(u(x)) = \int u(\lambda)\, m[x(d\lambda)]$. We use the notation $M_x = \{m \in M : m(x)$ exists and is finite$\}$.

Theorem 4.16. *An observable x is bounded if and only if $M_x = M$.*

Proof. Necessity is trivial. To prove sufficiency suppose x is unbounded. Then there are distinct numbers $\lambda_n \in \sigma(x)$ such that $|\lambda_n| > 2^{n+1}$, $n = 1, 2, \ldots$. Let E_n be a sequence of disjoint open intervals of diameter less than one centered at λ_n and let $a_n = x(E_n) \neq 0$. Now there exist $m_j \in M$ such that $m_j(a_j) = 1$. Since M is closed under mixtures $m = \sum 2^{-j} m_j \in M$. Since $a_i \perp a_j$, $m_i(a_j) = 0$ for $i \neq j$ and $m(a_j) = 2^{-j}$. Now suppose $m(x)$ exists and hence $\int |\lambda|\, m[x(d\lambda)]$ exists. But $\int |\lambda|\, m[x(d\lambda)] \geq \sum \int_{E_i} |\lambda|\, m[x(d\lambda)] \geq \sum (2^{i+1} - 1)2^{-i} = +\infty$. This contradiction shows that $m(x)$ does not exist and finishes the proof.

The expectations of an observable in different states are, physically, values obtained for the observable using a macroscopic experiment. These values are, in a sense, averages over the microscopic or "actual" values of the observable which are the numbers in the spectrum. Thus one would expect the expectations to "smooth out" or "fill in" the spectral values. This is shown in the next theorem whose proof may be found in [13].

Theorem 4.17. *The set* $\{m(x): m \in M_x\}$ *is the smallest closed interval containing* $\sigma(x)$.

Even though the states separate propositions it is not known if the expectations separate bounded observables. (This question will be treated in the next subsection.) The logic (L, M) satisfies *condition* U (uniqueness) if $m(x) = m(y)$ for all $m \in M$ implies the bounded observables x and y are equal.

We next prove a spectral mapping theorem.

Theorem 4.18. *Let x be an observable, f, g Borel functions and suppose $E \in \mathscr{B}(R)$ satisfies $x(E) = 1$. (i) If $f(\lambda) = g(\lambda)$ for all $\lambda \in E$, then $f(x) = g(x)$. (ii) $\sigma(f(x)) \subset \mathrm{Cl}\, f(E)$. (iii) If f is continuous $\sigma(f(x)) = \mathrm{Cl}\, f(\sigma(x))$. If f is continuous and x is bounded $\sigma(f(x)) = f(\sigma(x))$. (iv) In general $\sigma(f(x)) = \bigcap\{\mathrm{Cl}\, f(E): x(E) = 1\}$.*

Proof. (i) $f(x)(F) = x(f^{-1}(F)) = x(E \cap f^{-1}(F)) = x(E \cap g^{-1}(F)) = g(x)(F)$. (ii) We first show $\sigma(f(x)) \subset \mathrm{Cl}\, f(R)$. Suppose $\lambda \in \sigma(f(x))$. Then if $\lambda \in U$ and U is open, we have $0 \neq f(x)(U) = x(f^{-1}(U))$. Therefore $f^{-1}(U) \neq \varnothing$ and $U \cap f(R) \neq \varnothing$. Thus $\lambda \in \mathrm{Cl}\, f(R)$ and $\sigma(f(x)) \subset \mathrm{Cl}\, f(R)$. Now let $\lambda_0 \in f(E)$ and define $g(\lambda) = f(\lambda)$ if $\lambda \in E$, and $g(\lambda) = \lambda_0$ otherwise. By part (i) we have $\sigma(f(x)) = \sigma(g(x)) \subset \mathrm{Cl}\, g(R) \subset \mathrm{Cl}\, f(E)$. (iii) Letting $E = \sigma(f(x))'$ we have $f(x)(E) = 0$ and thus $x(f^{-1}(E)) = 0$. Since f is continuous the set $f^{-1}(E)$ is open and hence in $\sigma(x)'$. Therefore $f^{-1}(E) \cap \sigma(x) = \varnothing$ and $E \cap f(\sigma(x)) = \varnothing$. Hence $f(\sigma(x)) \subset \sigma(f(x))$ and $\mathrm{Cl}\, f(\sigma(x)) \subset \sigma(f(x))$ since the latter is closed. The inclusion in the other direction follows from (ii). (iv) Suppose $\lambda \in \bigcap\{\mathrm{Cl}\, f(E): x(E) = 1\}$ and $\lambda \in U$ where U is open and $f(x)(U) = 0$. Then $1 = f(x(U')) = x(f^{-1}(U'))$. Therefore $\lambda \in \mathrm{Cl}\, f[f^{-1}(U')] \subset U'$ since the latter is closed. But this is a contradiction, and hence $\lambda \in \sigma(f(x))$. The inclusion in the other direction follows from (ii).

Theorem 4.19. *(i) If x_1 and x_2 are compatible observables one of which is bounded then $\sigma(x_1 + x_2) \subset \sigma(x_1) + \sigma(x_2)$ and $\sigma(x_1 x_2) \subset \sigma(x_1)\sigma(x_2)$. (ii) If $x_1 \leftrightarrow x_2$ are bounded observables, then $|x_1 + x_2| \leq |x_1| + |x_2|$ and $|x_1 x_2| \leq |x_1|\,|x_2|$.*

Proof. There exists an observable x and Borel functions u_1, u_2 such that $x_1 = u_1(x)$, $x_2 = u_2(x)$. Suppose x_1 is bounded. We shall indicate how the proof proceeds by considering the special case in which u_1 and

u_2 are continuous. In the general case, one must contend with some technical set theoretic complications. Using Theorem 4.15, Theorem 4.18 (iii) and the fact that $\sigma(x_1)$ is compact we have

$$
\begin{aligned}
\sigma(x_1 + x_2) &= \sigma(u_1(x) + u_2(x)) = \sigma((u_1 + u_2)(x)) \\
&= \text{Cl}\,(u_1 + u_2)(\sigma(x)) \subset \text{Cl}\,(u_1(\sigma(x)) + u_2(\sigma(x))) \\
&= \text{Cl}\,u_1(\sigma(x)) + \text{Cl}\,u_2(\sigma(x)) \\
&= \sigma(u_1(x)) + \sigma(u_2(x)) = \sigma(x_1) + \sigma(x_2).
\end{aligned}
$$

The rest of (i) is similar and (ii) follows easily.

A set of compatible observables K is *complete* if $x \leftrightarrow K$ implies $x \in K$. Complete sets of compatible observables are important in certain physical situations in which the state of the system is completely described by finding values for these observables. An easy Zorn's lemma argument shows that every nonempty set of compatible observables is contained in a complete set.

Theorem 4.20. *A complete set K of bounded observables in a logic satisfying condition C is a commutative real Banach algebra with unity satisfying*:

(i) $|x^2| = |x|^2$ *for all $x \in K$*
(ii) x^2 *is a continuous function of x*
(iii) $|x^2 - y^2| \leq \max(|x^2|, |y^2|)$.

Proof. Define the additive and multiplicative identity $0, I$ as the unique observables with spectrum $\{0\}$ and $\{1\}$, respectively. It easily follows that K is a commutative normed algebra with unity. It follows from the spectral mapping theorem that $\sigma(x^2) = \sigma(x)^2$ so (i) follows. If $x_i \to x$ in norm, then $|x_i|$ is uniformly bounded and $|x^2 - x_i^2| = |(x - x_i)(x + x_i)| \leq (|x| + |x_i|)|x - x_i|$ so $x_i^2 \to x^2$ and (ii) follows. Now suppose $\lambda \in \sigma(x^2 - y^2)$. By Theorem 4.18 and Theorem 4.19 $\sigma(x^2 - y^2) \subset \sigma(x^2) - \sigma(y^2) = \sigma(x)^2 - \sigma(y)^2$. Thus $\lambda = \lambda_1 - \lambda_2$ where $\lambda_1 \in \sigma(x)^2$, $\lambda_2 \in \sigma(y)^2$. If $\lambda_1 - \lambda_2 \geq 0$, then $|\lambda| = \lambda \leq \lambda_1 \leq \max(\lambda_1, \lambda_2)$. If $\lambda_1 - \lambda_2 \leq 0$, then $|\lambda| = -\lambda \leq \lambda_2 \leq \max(\lambda_1, \lambda_2)$. (iii) now follows. We next show that K is complete. Let x_n be a Cauchy sequence in K, and let B be the smallest Boolean sub σ-algebra containing $\bigcup R(x_n)$. Since B is separable there is an observable z such that $R(z) = B$. We now show $z \in K$. Otherwise there is an $x \in K$ and $x \leftrightarrow z$. But since $x \leftrightarrow x_n$ there is a Boolean sub σ-algebra B_1 which contains $R(x) \cup (\bigcup R(x_n))$. But then B_1 contains

$\bigcup R(x_n)$ but cannot contain B which contradicts the minimality of B so $z \in K$. Now there exist Borel functions u_n such that $x_n = u_n(z)$ and since x_n is Cauchy there are positive integers $n(p)$, $p = 1, 2, \ldots$, such that $n, m \geq n(p)$ implies $|u_n(z) - u_m(z)| \leq p^{-1}$. Letting $\Delta(\varepsilon) = \{\lambda: |\lambda| \leq \varepsilon\}$ we have $\sigma[u_n(z) - u_m(z)] \subset \Delta(p^{-1})$ and $0 = [u_n(z) - u_m(z)](\Delta(p^{-1})') = z\{w: |u_n(w) - u_m(w)| > p^{-1}\}$ for $n, m \geq n(p)$. Letting

$$N(p) = \bigcup_{n,m \geq n(p)} \{w: |u_n(w) - u_m(w)| > p^{-1}\}$$

we have $|u_n(w) - u_m(w)| \leq p^{-1}$, $n, m \geq n(p)$, on $N(p)'$ and $z[N(p)] = 0$. Now if $N = \bigcup N(p)$, then $z(N) = \bigvee z[N(p)] = 0$. We assert that u_n is uniformly Cauchy on N'. Indeed, if $\varepsilon > 0$ then there is an integer q such that $q^{-1} < \varepsilon$ and if $n, m > n(q)$ we have $|u_n(w) - u_m(w)| < q^{-1} < \varepsilon$ on $N(q)'$ and hence on N'. Therefore u_n converges uniformly on N' to a Borel function u. We now show that $u_n(z) \to u(z)$. For any $\varepsilon > 0$ if n is sufficiently large we have $\{w: |u_n(w) - u(w)| > \varepsilon\} \subset N$. Hence $[u_n(z) - u(z)](\Delta(\varepsilon)') = z\{w: u_n(w) - u(w) \in \Delta(\varepsilon)'\} = z\{w: |u_n(w) - u(w)| > \varepsilon\} = 0$. So for n sufficiently large $\sigma[u_n(z) - u(z)] \subset \Delta(\varepsilon)$ and $|u_n(z) - u(z)| < \varepsilon$. Since $z \in K$ we have, of course, that $u(z) \in K$ and thus $x_n \to u(z) \in K$ and K is complete.

Corollary 4.21. *A complete set of bounded observables in a logic satisfying condition C is isometrically isomorphic to the continuous real valued functions on a compact Hausdorff space.*

Proof. This follows from a theorem due to Segal [47, p. 935].

Theorem 4.22. *Let K be a complete set of bounded observables on a logic satisfying condition C. Then the map $m: K \to R$ defined by $x \to m(x)$ is a bounded linear functional on K and $|x| = \sup\{|m(x)|: m \in M\}$.*

Proof. Suppose $x, y \in K$ and $x = u_1(z)$, $y = u_2(z)$. Then

$$m(x + y) = m[u_1(z) + u_2(z)] = \int [u_1(\lambda) + u_2(\lambda)]m[z(d\lambda)]$$
$$= \int u_1(\lambda)m[z(d\lambda)] + \int u_2(\lambda)m[z(d\lambda)] = m(x) + m(y);$$
$$m(\alpha x) = \int \alpha\lambda m[x(d\lambda)] = \alpha m(x).$$

The following proves boundedness: $|m(x)| = \left|\int_{\sigma(x)} \lambda m[x(d\lambda)]\right| \leq \int_{\sigma(x)} |\lambda| \, m[x(d\lambda)] \leq |x|$. Hence $\sup\{|m(x)|: m \in M\} \leq |x|$. Let $\varepsilon > 0$ be given, let $\lambda_0 \in \sigma(x)$, and define $E = \{\omega: |\omega - \lambda_0| < \varepsilon\}$. Since E is open

$x(E) \neq 0$ and hence there is an $m \in M$ such that $m[x(E)] = 1$. Therefore $|m(x)| = \left|\int_{\sigma(x)} \lambda m[x(d\lambda)]\right| \geq |\lambda_0| - \varepsilon$ and

$$|\lambda_0| \leq |m(x)| + \varepsilon \leq \sup\{|m(x)| : m \in M\} + \varepsilon.$$

Thus $\sup\{|\lambda| : \lambda \in \sigma(x)\} \leq \sup\{|m(x)| : m \in M\} + \varepsilon$ and since $\varepsilon > 0$ was arbitrary $|x| \leq \sup\{|m(x)| : m \in M\}$.

Let X denote the collection of bounded observables on a logic (L, M) satisfying condition C. Although multiplication by real numbers is defined for all $x \in X$ we can only define addition for compatible elements; thus X is not a linear space. Because of this fact some strange things can happen. For example, even though a Cauchy sequence of compatible observables must converge, there is no guarantee that the limit is unique. Thus it is possible for a sequence of bounded observables to converge to two different bounded observables. If (L, M) satisfies condition U however, this cannot happen.

Corollary 4.23. *If a sequence of compatible observables x_i converges to a bounded observable x, then $m(x_i) \to m(x)$ for all $m \in M$. If (L, M) satisfies condition U then x is unique.*

Proof. By Theorem 4.22 $|m(x_i) - m(x)| = |m(x_i - x)| \leq |x_i - x|$ so $m(x_i) \to m(x)$. If $x_i \to y$ then $m(x_i) \to m(y)$. Thus $m(x) = m(y)$ and $x = y$ follows from condition U.

Let us now view M as a subset of all the bounded linear functionals X' on X. A functional f on X is said to be *linear* if $x \leftrightarrow y$ implies $f(\alpha x + \beta y) = \alpha f(x) + \beta f(y)$. f is *bounded* if $|f| = \sup\{|f(x)| : |x| \leq 1\} < \infty$. It is straightforward to show that $|f| = \inf\{N : |f(x)| \leq N|x|, x \in X\}$, in particular $|f(x)| \leq |f||x|$. We define $(\alpha f)(x) = \alpha f(x)$ and $(f + g)(x) = f(x) + g(x)$. These are both bounded linear functionals if f, g are and $|\alpha f| = |\alpha||f|$, $|f + g| \leq |f| + |g|$. We thus see that X' is a normed linear space even though X is not. In fact X' is a Banach space and the proof is almost identical to the case when X is linear space [7, p. 61]. We now endow X' with a weak topology τ (usually called the weak $*$ topology). A neighborhood basis for τ of the origin will be sets of the form $N(\varepsilon, x_1, \ldots, x_n) = \{f \in X' : |f(x_i)| < \varepsilon, i = 1, 2, \ldots, n\}$. Now τ makes X' a locally convex space. It can be shown that a net of states m_α converges to a state m in the τ topology if and only if $m_\alpha(a) \to m(a)$ for every proposition a. The proof of the next theorem may be found in [13].

Theorem 4.24. *If X' with the τ topology is metrizable, then M is a compact subset of X' in the τ topology.*

One can give examples to show that this theorem does not hold if X' is not metrizable in the τ topology.

Corollary 4.25. *If X' is metrizable in the τ topology then M is the τ-closed convex hull of its pure states.*

Proof. The pure states are the extreme points of M and thus the Corollary follows from Theorem 4.24 and the Krein–Milman theorem.

We have already postulated that L has a full set of states. However, this does not *a priori* guarantee that there are any pure states. Since the pure states are those in which we have a maximum amount of information concerning the condition of the system it is important to show that there are a lot of pure states.

Corollary 4.26. *Suppose X' is metrizable in the τ topology.* (i) *If $a \neq b$ there is a pure state $m \in M$ such that $m(a) \neq m(b)$.* (ii) *If $a \neq 0$ there is a pure state m such that $m(a) = 1$.*

Proof.[1] (i) Suppose $m(a) = m(b)$ for every pure state m. Then convex combinations of pure states agree on a and b and so do limits of sequences of convex combinations. It follows from Corollary 4.25 that all states agree on a and b which is a contradiction. (ii) Let $M_a = \{m \in M : m(a) = 1\}$. Then M_a is a nonempty subset of M which is closed in the τ topology. Thus M_a is compact and is therefore the closed convex hull of its extreme points. Let m_0 be an extreme point of M_a. To show m_0 is an extreme point of M suppose $m_0 = \lambda m_1 + (1 - \lambda)m_2$ for $m_1, m_2 \in M$, $0 < \lambda < 1$. Then $1 = m_0(a) = \lambda m_1(a) + (1 - \lambda)m_2(a)$. Hence $m_1(a) = m_2(a) = 1$ and $m_1, m_2 \in M_a$. Thus $m_1 = m_2 = m_0$.

D. UNIQUENESS AND EXISTENCE

Recall that the uniqueness condition U states that if the expectations of two bounded observables are equal in every state, then the observables are equal. We next consider an existence property concerning the existence of the sum of two bounded observables. Although we cannot define the sum of two bounded observables in a direct way, we can define it indirectly as follows. We say that an observable z is the *sum* of two bounded observables x and y if $m(z) = m(x) + m(y)$ for every state m.

---

[1] The author is indebted to Harry Mullikin for the proof of (ii) in the above corollary.

If the sum of any two bounded observables exists we say that the logic (L, M) satisfies *condition* E. This existence property is so important that it is *postulated* in some models for quantum mechanics (cf. [46, 58]). In particular, Segal postulates that the (bounded) observables form a Banach space in which squares exist and which satisfies (i), (ii), and (iii) of Theorem 4.20. We may now pose the question as to whether an arbitrary logic (L, M) satisfies conditions U and E. In this subsection we give a counterexample to show that (L, M) need not satisfy condition E. Although condition U is unanswered in general, we show that it is satisfied in certain cases.

Let us consider two examples which illustrate our problem. Let (Ω, F) be a measurable space. As we have seen, the states of this system are the probability measures on Ω, and the observables are essentially the random variables on F. Suppose M is the set of all probability measures on F. The uniqueness condition may be stated as follows: If f and g are bounded measurable functions, does $\int f\,d\mu = \int g\,d\mu$ for every probability measure imply $f = g$? The answer, which is yes, may be seen as follows. Let $p \in \Omega$ and let μ be a probability measure concentrated at p. Now it is easily seen that $\int_\Lambda f\,d\mu = \int_\Lambda g\,d\mu$ for every $\Lambda \in F$. Therefore $f = g$ a.e. with respect to μ and in particular $f(p) = g(p)$. The existence condition holds even more trivially since $f + g$ is a bounded measurable function and $\int (f + g)\,d\mu = \int f\,d\mu + \int g\,d\mu$ for every probability measure μ. For our next example consider Example 2 of Section IV.A. The uniqueness condition becomes: If A and B are bounded self-adjoint operators, does $\langle \phi, A\phi \rangle = \langle \phi, B\phi \rangle$ for all $\phi \in H$, $\|\phi\| = 1$ imply $A = B$? It is well-known that the answer is yes. Condition E is again trivially satisfied since $A + B$ is a bounded self-adjoint operator and $\langle \phi, (A + B)\phi \rangle = \langle \phi, A\phi \rangle + \langle \phi, B\phi \rangle$ for every $\phi \in H$.

We now consider condition U more closely.

Theorem 4.27. *(i) If $m[u(x)] = m[u(y)]$ for every Borel function u and every $m \in M$, then $x = y$. (ii) If $x \leftrightarrow y$ and $m(x) = m(y)$ for every $m \in M$ then $x = y$. (iii) If x or y has a one or two point spectrum and $m(x) = m(y)$ for every $m \in M$, then $x = y$.*

Proof. (i) Let $E \in \mathscr{B}(R)$ and let χ_E denote its characteristic function. Then for every $m \in M$, $m[x(E)] = m[\chi_E(x)] = m[\chi_E(y)] = m[y(E)]$. Thus $x(E) = y(E)$ and $x = y$. (ii) Since $x \leftrightarrow y$, $m(x - y) = m(x) - m(y) = 0$. Applying Theorem 4.22, $|x - y| = 0$. Thus $x - y = 0$ and $x = y$.

(iii) Since $\{m(x): m \in M\} = \{m(y): m \in M\}$ it follows from Theorem 4.17 that $\sigma(x) = \sigma(y)$. Hence, if either x or y has a one point spectrum, $x = y$. Now suppose both x and y have a two point spectrum $\{\lambda_1, \lambda_2\}$. Then for every $m \in M$ we have $m(x) = \lambda_1 m[x(\lambda_1)] + \lambda_2 m[x(\lambda_2)] = m(y) = \lambda_1 m[y(\lambda_1)] + \lambda_2 m[y(\lambda_2)]$. Since $x(\lambda_1) + x(\lambda_2) = y(\lambda_1) + y(\lambda_2) = 1$ we have $m[x(\lambda_1)](\lambda_1 - \lambda_2) = m[y(\lambda_1)](\lambda_1 - \lambda_2)$ and $m[x(\lambda_1)] = m[y(\lambda_1)]$. Hence $x(\lambda_1) = y(\lambda_1)$ and $x = y$.

We say that a logic (L, M) is *quite full* in case $a \leq b$ if $m(b) = 1$ whenever $m(a) = 1$. It can be shown that if a set of states M is closed under convex combinations and satisfies the above property then M is a full set of states on L.

Lemma 4.28. *Let x and y be bounded observables on a quite full logic (L, M) and suppose $m(x) = m(y)$ for all $m \in M$. Then $\lambda_0 = \max\{\lambda: \lambda \in \sigma(x)\} = \max\{\lambda: \lambda \in \sigma(y)\}$ and $x(\lambda_0) = y(\lambda_0)$.*

Proof. That $\max\{\lambda: \lambda \in \sigma(x)\} = \max\{\lambda: \lambda \in \sigma(y)\}$ follows from Theorem 4.17. Now suppose $m[x(\lambda_0)] = 1$ and $m[y(\lambda_0)] \neq 1$. Then there is a number $\mu < \lambda_0$ such that $m[y(-\infty, \mu])] > 0$. We now have

$$\lambda_0 = m(x) = m(y) = \int_{(-\infty, \lambda_0]} \lambda m[y(d\lambda)]$$

$$= \left(\int_{(-\infty, \mu)} + \int_{[\mu, \lambda_0]}\right)\lambda m[y(d\lambda)]$$

$$\leq \mu m[y(-\infty, \mu)] + \lambda_0 m(y[\mu, \lambda_0]) < \lambda_0,$$

which is a contradiction. Thus $m[y(\lambda_0)] = 1$ whenever $m[x(\lambda_0)] = 1$ and hence $x(\lambda_0) \leq y(\lambda_0)$. By symmetry $x(\lambda_0) = y(\lambda_0)$.

Theorem 4.29. *Let x and y be bounded observables on a quite full logic (L, M), and suppose $\sigma(x)$ has at most one limit point. If $m(x) = m(y)$ for every $m \in M$ then $x = y$.*

Proof. The most general such x has a point $\lambda_0 \in \sigma(x)$ which is a limit point from both above and below of elements of $\sigma(x)$. The other cases will follow in a similar manner. We can also assume without loss of generality that $\lambda_0 = 0$. Let the points of $\sigma(x)$ be ordered as follows: $\mu_1 < \mu_2 < \cdots < \lambda_0 < \cdots < \lambda_2 < \lambda_1$. Now by Lemma 4.28 $\max\{\lambda: \lambda \in \sigma(y)\} = \lambda_1$ and $y(\lambda_1) = x(\lambda_1)$. Denoting the characteristic function of $\{\lambda_1\}$ by χ_{λ_1} let $x_1 = x - \lambda_1\chi_{\lambda_1}(x)$ and $y_1 = y - \lambda_1\chi_{\lambda_1}(y)$. Letting f be the identity function $f(\lambda) = \lambda$ we have for $E \in \mathcal{B}(R)$,

$$x_1(E) = (f - \lambda_1\chi_{\lambda_1})(x)(E)$$

$$= x[(f - \lambda_1\chi_{\lambda_1})^{-1}(E)]$$

$$= \begin{cases} x(E) \wedge x(\lambda_1)', & \text{if} \quad 0 \notin E, \\ x(E) \vee x(\lambda_1), & \text{if} \quad 0 \in E. \end{cases} \tag{4.1}$$

It is now easy to see that $\sigma(x_1) = \sigma(x) \cap \{\lambda_1\}'$; $x_1(\lambda_i) = x(\lambda_i)$, $i = 2, 3,$ ...; and that $x_1(\mu_i) = x(\mu_i)$, $i = 1, 2, \dots$. Now $m(x_1) = m(x) - \lambda_1 m[x(\lambda_1)]$ $= m(y) - \lambda_1 m[y(\lambda_1)] = m(y_1)$. Applying Lemma 4.28

$$\lambda_2 = \max\{\lambda: \lambda \in \lambda \in \sigma(y_1)\}$$

and $y_1(\lambda_2) = x_1(\lambda_2) = x(\lambda_2)$. It now follows by applying Eq. (4.1) to y_1 and y that λ_2 is the second largest number in $\sigma(y)$ and $y(\lambda_2) = y_1(\lambda_2) = x(\lambda_2)$. Continuing this process with the λ_i's and also the μ_i's we have $\{\lambda_i, \mu_i: i = 1, 2, \dots\} \subset \sigma(y)$ and $y(\lambda_i) = x(\lambda_i)$, $y(\mu_i) = x(\mu_i)$, $i = 1, 2, \dots$. Since λ_0 is a limit point of the λ_i's it follows that $\lambda_0 \in \sigma(y)$, $\{\lambda_i, \mu_i: i = 1, 2, \dots\} \subset \sigma(y)$ and $y(\lambda_0) = y(\{\lambda_i, \mu_i: i = 1, 2, \dots\}') = [\sum y(\lambda_i) + \sum y(\mu_i)]' = [\sum x(\lambda_i) + \sum x(\mu_i)]' = x(\lambda_0)$. Hence $y = x$.

We next consider condition E. First of all it is clear that if $x \leftrightarrow y$ then $x + y$ exists. We now construct an example of a logic in which the sum of no two noncompatible observables exists. An *antilattice* is a complemented lattice in which the supremum of any two nonzero elements is 1. It is easily seen that an antilattice together with its set of states forms a (quite full) logic.

Lemma 4.30. *A Boolean subalgebra of an antilattice L can have at most four elements.*

Proof. Let a, b be distinct elements of L not equal to 0 or 1 and $a \neq b'$. We now show that $a \leftrightarrow b$. Suppose, on the contrary, that $a = a_1 + c$ and $b = b_1 + c$ where $a_1 \perp b_1$. Now neither a_1 nor b_1 is 0, hence $c = 0$, since otherwise $a_1 + c = 1$. Therefore $a \perp b$. But this is impossible since then $b' = a \vee b' = 1$. The result follows since the elements of a Boolean subalgebra must be compatible.

Corollary 4.31. *Every observable on an antilattice is of the form* $x = (\lambda_1 - \lambda_2)\chi_{\lambda_1}(x) + \lambda_2 I$, *where* $\sigma(x) = \{\lambda_1, \lambda_2\}$.

Proof. It follows from Lemma 4.30 that if x is an observable then x has one or two point spectrum so suppose $\sigma(x) = \{\lambda_1, \lambda_2\}$. (If x has one point in its spectrum we let $\lambda_1 = \lambda_2$.) Thus x may be written $x = \lambda_1 \chi_{\lambda_1}(x) + \lambda_2 \chi_{\lambda_2}(x)$. Since $I = \chi_{\lambda_1}(x) + \chi_{\lambda_2}(x)$ the corollary follows.

It is interesting that since every observable on an antilattice has one or two point spectrum, condition U holds by Theorem 4.27.

Theorem 4.32. *If x and y are noncompatible observables on an anti-lattice, then $x + y$ does not exist.*

Proof. Suppose $z = x + y$ exists. Applying Corollary 4.31 we have $z = x + y = (\lambda_1 - \lambda_2)\chi_{\lambda_1}(x) + (\mu_1 - \mu_2)\chi_{\mu_1}(y) + (\lambda_2 + \mu_2)I$, where $\sigma(x) = \{\lambda_1, \lambda_2\}$, $\sigma(y) = \{\mu_1, \mu_2\}$. Since $(\lambda_2 + \mu_2)I$ is compatible with every observable it follows that $z_0 = \chi_{\lambda_1}(x) + \beta \chi_{\mu_1}(y)$ must exist, where $\beta = (\mu_1 - \mu_2)/(\lambda_1 - \lambda_2)$ (notice $\lambda_1 \neq \lambda_2$). Now if $\sigma(z_0) = \{\omega_1, \omega_2\}$ we have $(\omega_1 - \omega_2)\chi_{\omega_1}(z_0) + \omega_2 I = \chi_{\lambda_1}(x) + \beta \chi_{\mu_1}(y)$. Hence

$$(\omega_1 - \omega_2)m[z_0(\omega_1)] + \omega_2 = m[x(\lambda_1)] + \beta m[y(\mu_1)]$$

for every $m \in M$. Now it is easily seen that on an antilattice L, given $0 < \alpha < 1$ and $c \in L$, there is a state m such that $m(c) = \alpha$ and $m(d) = 0$ where d is any proposition not equal to c, c', or 1. Now suppose $z_0(\omega_1) = x(\lambda_1)$. Letting m be a state which is 0 on $y(\mu_1)$ and 1 on $x(\lambda_1)$ we have $\omega_1 = 1$. Letting m be a state which is 0 on $x(\lambda_1)$ and 1 on $y(\mu_1)$ we have $\omega_2 = \beta$ and hence $m[x(\lambda_1)] + m[y(\mu_1)] = 1$ for all $m \in M$ which is impossible. Hence $z_0(\omega_1) \neq x(\lambda_1)$, and in a similar way $z_0(\omega_1) \neq x(\lambda_1)'$, $y(\mu_1)$, $y(\mu_1)'$, 0, 1. Now let m satisfy $m[x(\lambda_1)] = \alpha \neq \omega_2$ and $m[y(\mu_1)] = m[z_0(\omega_1)] = 0$. This gives a contradiction and hence z_0 does not exist.

One can also give examples of logics on which the sum of certain noncompatible observables exists while the others do not. If it turns out that the uniqueness condition does not necessarily hold on a logic (L, M) then even if the sum of two bounded observables exists it may not be unique. For example, suppose x and y are distinct bounded observables and $m(x) = m(y)$ for all $m \in M$. Now $2x = x + y$ and $2y = x + y$ is clear. Thus $x + y$ exists but is not unique.

We complete this section by showing that a quite full logic satisfying condition E is a lattice. If $a \in L$ we denote by x_a the unique observable which satisfies $x_a(1) = a$, $x_a(0) = a'$. We call x_a the *characteristic observable* for a. Notice that $\chi_E(y) = x_{y(E)}$ for any observable y, $E \in \mathscr{B}(R)$.

Theorem 4.33. *Let* (L, M) *be a quite full logic which satisfies condition E. Then L is a lattice and in fact* $a \wedge b = (x_a + x_b)(\{2\})$. *If* $m(a) = m(b) = 1$, *then* $m(a \wedge b) = 1$.

Proof. Let $z = x_a + x_b$. Notice that $0 \le m(z) \le 2$ for every $m \in M$ and hence by Theorem 4.17, $\sigma(z) \subset [0, 2]$. Let $c = z(\{2\})$ and suppose $m(c) = 1$. Then $m(z) = 2$ and $m(a) = m(b) = 1$. Hence $c \le a$, $c \le b$. Now suppose $d \le a$, $d \le b$. Then $m(d) = 1$ implies $m(z) = 2$ which implies $m(c) = 1$. Thus $d \le c$ and $c = a \wedge b$. For the last statement suppose $m(a) = m(b) = 1$. Then $m(x_a + x_b) = m(a) + m(b) = 2$ and hence $1 = m[(x_a + x_b)(\{2\})] = m(a \wedge b)$.

Corollary 4.34. *If* (L, M) *is a quite full logic which satisfies condition E, then*

(i) $(x_a + x_b)(\{0\}) = a' \wedge b'$;
(ii) $(x_a - x_b)(\{1\}) = a \wedge b'$; *and*
(iii) *if* $a, b, \ldots, c$ *are n elements of L,*

$$(x_a + x_b + \cdots + x_c)(\{n\}) = a \wedge b \wedge \cdots \wedge c.$$

This subsection is based on [14] where further details may be found.

E. HIDDEN VARIABLES

The question of hidden variables in quantum mechanics is, in a sense, a philosophical question which asks whether probability is really necessary in quantum theory. The situation, according to the hidden variable proponents, is similar to that occurring in statistical mechanics. If we have a large number of molecules in a container, then it is impractical, and in fact nearly impossible, to describe the motion of each particle individually. For this reason one gives a macroscopic description of the system using macroscopic quantities such as volume, temperature, and pressure. In terms of these macroscopic quantities one can make only statistical predictions about the motions of the individual particles. Thus probabilities enter only because of human deficiencies and not because of any intrinsic, inherent properties of the system. If one could describe the condition of the system more precisely, then theoretically the probabilistic nature of the system would vanish. According to some physicists and philosophers this could be the case in quantum mechanics. It is possible that there are certain "hidden variables," as yet unknown, which would determine the condition of a quantum mechanical system so precisely that statistical considerations would vanish and exact phenomenological predictions could be made. Once the values of these hidden variables are known one could verify or refute any experimental proposition concerning the system with certainty instead of merely giving probabilities for their truth or falsity. In our axiomatic system the existence of hidden variables would imply the existence of "dispersion-free states," that is, states which have only the values 0 or 1.

Hidden variable models have been given for certain simple quantum mechanical systems [5, 6] giving "exact" predictions and thus dispensing entirely with statistical considerations. It thus appears that the existence of hidden variables depends upon the model used for describing the physical system. The question thus arises as to whether the axiomatic model considered here admits hidden variables. Von Neumann [59] has shown that the usual Hilbert space model for quantum mechanics does not admit hidden variables, while Jauch and Piron [22] have shown that certain more general models do not. In this section we show that the present model, which is more general than that of Jauch and Piron, does not admit hidden variables, thus forcing us to contend with statistical considerations. The contents of this subsection are based on [17].

A state m on a logic (L, M) is *dispersion-free* if m has only the values 0 and 1 and satisfies:

(D-1) if $m(a) = m(b) = 1$ and if $a \wedge b$ exists, then $m(a \wedge b) = 1$;

(D-2) if $\{a_\alpha : \alpha \in A\}$ are mutually compatible, $\bigwedge a_\alpha$ exists, and $m(a_\alpha) = 1$, $\alpha \in A$, then $m(\bigwedge a_\alpha) = 1$.

The logic (L, M) is *compatibly complete* if for any compatible set $\{a_\alpha : \alpha \in A\}$, $\bigwedge a_\alpha$ exists; and (L, M) is *complete* if $\bigwedge b_\alpha$ exists for any collection of propositions b_α. (L, M) *admits hidden variables* if it is compatibly complete and if for any nonzero $a \in L$, there is a dispersion-free state m such that $m(a) = 1$. The center Z of L is the set of propositions which are compatible with every proposition. Notice that L is compatible if and only if $L = Z$. We say that L is *coherent* or *trivial* if $Z = \{0, 1\}$ or $L = \{0, 1\}$, respectively. A proposition a is an *atom* if $a \neq 0$ and if $b \leq a$ implies b is a or 0. L is *atomic* if every nonzero proposition contains an atom.

Theorem 4.35. *A compatibly complete logic (L, M) has a dispersion-free state if and only if L has an atom in its center.*

Proof. To prove necessity, let m be a dispersion-free state and let $L_m = \{a \in L : m(a) = 1\}$. Note that $L_m \neq \emptyset$ and let T be a totally ordered subset of L_m. Then since the elements of T are compatible $a_0 = \bigwedge \{a \in T\}$ exists and by (D-2) $m(a_0) = 1$. Thus $a_0 \in L_m$ and $a_0 \leq a$ for all $a \in T$. By Zorn's lemma L_m has a minimal element a_1. We now show that a_1 is a lower bound for L_m (i.e., $a_1 \leq a$ for all $a \in L_m$). Let $a_2 \in L_m$. Now there is a nonzero $a_3 \in L$ such that $a_3 \leq a_1$, a_2 since otherwise $a_1 \wedge a_2 = 0$ and $m(a_1 \wedge a_2) = 0$ which contradicts (D-1). Now suppose $m(a_3) = 0$. Then $m(a_1 - a_3) = 1$ and $a_1 - a_3 \in L_m$. Since $a_1 - a_3 \leq a_1$, $a_1 - a_3 = a_1$, and hence $a_3 = a_3 \wedge a_1 = 0$ which is a contradiction. Thus $m(a_3) = 1$, $a_3 \in L_m$, and hence $a_3 = a_1$. Therefore $a_1 \leq a_2$ and a_1 is a lower bound for L_m. We next show that $a_1 \in Z$. If $b \in L_m$, then $a_1 \leq b$ and $a_1 \leftrightarrow b$. If $b \notin L_m$, then $b' \in L_m$ and $a_1 \leq b'$. Thus $a_1 \leftrightarrow b'$ and it follows that $a_1 \leftrightarrow b$. To show a_1 is an atom notice $a_1 \neq 0$ and suppose $b \leq a_1$. If $b \in L_m$ then $b = a_1$. If $b \notin L_m$ then $b' \in L_m$ and $a_1 \leq b'$. Thus $b \leq b'$ and hence $b = b \wedge b \leq b \wedge b' = 0$. To prove sufficiency, let $a_1 \in Z$ be an atom and let $b \in L$. We show that either $a_1 \leq b$ or $a_1 \leq b'$. Since $b \leftrightarrow a_1$, $a_1 \wedge b$ exists and $a_1 \wedge b \leq a_1$. Therefore either $a_1 \wedge b = a_1$ and $a_1 \leq b$ or $a_1 \wedge b = 0$ which implies $a_1 \leq b'$. Now let m be a state satisfying $m(a_1) = 1$. Then either $a_1 \leq b$ and $m(b) = 1$ or $b \leq a_1'$ and $m(b) = 0$. Thus m is dispersion-free.

Corollary 4.36 *A non trivial coherent compatibly complete logic (L, M) has no dispersion-free states.*

Proof. Suppose (L, M) has a dispersion-free state. Then by Theorem 4.35 there is an atom in the center $Z = \{0, 1\}$. Since 0 is not an atom, 1 must be an atom. But then L is trivial which is a contradiction.

Theorem 4.37. *A logic (L, M) admits hidden variables if and only if L is complete, atomic, and compatible.*

Proof. If $0 \neq a \in L$, then there is a dispersion-free state m such that $m(a) = 1$. By the proof of Theorem 4.35 there is an atom $a_1 \in Z$ such that $a_1 \leq a$. Let Z_1 be the collection of atoms in Z and let a, b be two propositions. Let

$$a_1 = \bigvee \{z \in Z_1 : z \leq a, z \leq b'\},$$
$$b_1 = \bigvee \{z \in Z_1 : z \leq b, z \leq a'\},$$
$$c = \bigvee \{z \in Z_1 : z \leq a, z \leq b\}.$$

Now a_1, b_1, c are mutually disjoint and $a = a_1 + c$, $b = b_1 + c$.

Hence L is atomic, compatible, and complete. To prove sufficiency, the states concentrated on atoms are dispersion-free.

Since we know from experiment that the logic describing a quantum mechanical system is not compatible, we see that the logic cannot admit hidden variables. We have shown that the admission of hidden variables is equivalent to L being complete, atomic, and compatible. But these properties are the essential features of a phase space description of the system. We thus see that (L, M) admits hidden variables if and only if (L, M) describes a classical system.

If we assume, as do Jauch and Piron, that L is a lattice, we get a much simpler proof under weaker conditions. We say that a logic (L, M) *admits hidden variables in the weak sense* if for any $0 \neq a \in L$ there is a state m with only the values 0 and 1 and satisfying (D-1) such that $m(a) = 1$.

Theorem 4.38. *If L is a lattice, and (L, M) admits hidden variables in the weak sense then L is compatible.*

Proof. Suppose (L, M) admits hidden variables in the weak sense, and $a, b \in L$. To show $a \leftrightarrow b$ it suffices by Corollary 4.3 to show that $b \leq (a - a \wedge b)' = a' + (a \wedge b)$ or that $c = b - [a' + (a \wedge b)] \wedge b = 0$. Let m be a state having only the values 0 and 1 and satisfying (D-1) for which $m(c) = 1$. Then $m(b) - m([a' + (a \wedge b)] \wedge b) = 1$. So $m(b) = 1$ and $m([a' + (a \wedge b)] \wedge b) = 0$. Hence $0 = m(a' + (a \wedge b)) = m(a') + m(a \wedge b)$ since otherwise we contradict (D-1). Therefore $m(a) = 1$ and $m(a \wedge b) = 0$, which contradicts (D-1). Thus $m(c) = 0$ and $c = 0$.

One can show by examples that a logic satisfying condition C need not be a lattice. For instance if $L = \{$subsets of $\{1, 2, 3, 4, 5, 6\}$ with an even number of elements$\}$ and M is the states on L, then (L, M) is such an example. Also one can show by examples that (D-1) is necessary for the above theorems to hold. For instance on an antilattice L_1 (D-1) does not hold. However L_1 is not compatible and except for (D-1), L_1 admits hidden variables.

V. A Generalized Probability Theory

We have seen in the last section that a generalized or noncommutative probability theory arises quite naturally from attempts to describe quantum mechanical phenomena. We have also seen from our hidden variable studies that in the present axiomatic formulation probabilistic considerations cannot be dispensed with. In this section we shall discuss some of the elementary consequences of this generalized theory. Much of our discussion will take place in the logic of closed subspaces of a Hilbert space which forms the usual setting in quantum mechanics. Hilbert space theory is usually used in one of two ways in conventional probability theory. It is used in the theory of second-order processes in which the random variables are square integrable, and it is used for processes almost all of whose sample functions are square integrable and the sample space may be taken as points of a Hilbert space. However, instead of representing the random variables as vectors in a Hilbert space as in the first case, they are represented by self-adjoint operators and instead of representing the events as Borel subsets of a Hilbert space as in the second case, they are represented by closed subspaces. Of course, new concepts appear in the generalized theory such as noncompatible propositions (events) and observables, so new techniques must be developed to handle these new situations.

Most of this section will be concerned with the notion of independence. Although the probabilistic aspects of this theory have been studied by both physicists and mathematicians [4, 37, 47, 48, 53, 55, 56], the notion of independence which is so basic to conventional probability has been largely neglected in the generalized theory. In this section *we shall assume that* (L, M) *is a quite full logic satisfying conditions* U *and* E. Conditions U and E are necessary if a comprehensive probability theory is to be developed. This is because many of the theorems of standard probability theory are concerned with sums of independent random variables, and to obtain an analogous generalized theory we must postulate the existence of unique sums. Notice that since condition E is assumed, L is a lattice (Theorem 4.33) and hence condition C holds (Lemma 4.5). The proof of the following lemma is straightforward.

Lemma 5.1. (i) L *is a* σ*-lattice, i.e., if* $a_1, a_2, \ldots \in L$, *then* $\bigvee a_i$ *exist. in* L. (ii) *if* $a_1 \leq a_2 \leq \cdots$ *and* $m \in M$, *then* $m(\bigvee a_i) = \lim m(a_i)$. *Also if* $a_1 \geq a_2 \geq \cdots$, *then* $m(\bigwedge a_i) = \lim m(a_i)$.

If $a_i \in L$ we define $\limsup a_i = \bigwedge_{k=1}^{\infty} (a_k \vee a_{k+1} \vee \cdots)$ and

$$\liminf a_i = \bigvee_{k=1}^{\infty} (a_k \wedge a_{k+1} \wedge \cdots).$$

Note that

$$\liminf a_i \leq \limsup a_i, \quad m(\limsup a_i) = \lim_{k \to \infty} (a_k \vee a_{k+1} \vee \cdots)$$

and

$$m(\liminf a_i) = \lim_{k \to \infty} m(a_k \wedge a_{k+1} \wedge \cdots)$$

for all $m \in M$. If lim sup $a_i =$ lim inf $a_i = a$, then we write $a =$ lim a_i.

Lemma 5.2. *If* lim $a_i = a$, *then* lim $m(a_i) = m(a)$ *for all* $m \in M$.

Proof. Since $m(a_k \vee a_{k+1} \vee \cdots) \geq \sup_{j \geq k} m(a_j)$ we have

$$\lim m(a_k \vee a_{k+1} \vee \cdots) \geq \lim \sup m(a_j).$$

Similarly

$$\lim m(a_k \wedge a_{k+1} \wedge \cdots) \leq \lim \inf m(a_j).$$

Therefore

$$\lim \sup m(a_j) \leq m(\lim \sup a_i) = m(a) = m(\lim \inf a_i) \leq \lim \inf m(a_j)$$

and hence lim $m(a_j) = m(a)$.

We shall show that the converse of this lemma does not always hold. The best we can do is the following.

Corollary 5.3. *If* lim $m(a_i) = m(a)$ *for every* $m \in M$, *then* lim inf $a_i \leq a \leq$ lim sup a_i.

Proof. $m(a) = \lim m(a_i) \geq \lim m(a_k \wedge a_{k+1} \wedge \cdots) = m(\lim \inf a_i)$ and $m(a) = \lim m(a_i) \leq \lim m(a_k \vee a_{k+1} \vee \cdots) = m(\lim \sup a_i)$. The corollary follows from the fact that (L, M) is quite full.

We now show that the converse of Lemma 5.2 holds in Example 1 (Section III.A) which is the usual measure-theoretic case. Suppose $m(a) = \lim m(a_i)$ for all $m \in M$ and yet lim inf $a_i < a <$ lim sup a_i. Let $p \in a$ and $p \notin$ lim inf a_i and let m be a probability measure concentrated at p. Now p is not in an infinite number of a_i's, hence there is a subsequence $a_{i(j)}$ of a_i such that $m(a_{i(j)}) = 0$. But then $1 = m(a) = \lim m(a_i) = \lim m(a_{i(j)}) = 0$, a contradiction. Thus lim inf $a_i = a$. Now suppose $p \in$ lim sup a_i, $p \notin a$ and again let m be concentrated at p. Then p is in an infinite number of the a_i's and hence $\lim(a_i) = 1$. But $m(a) = 0$ a contradiction and again lim sup $a_i = a$.

We next show that the converse of Lemma 5.2 does not necessarily hold in Example 2 (the Hilbert space example). For this purpose we give the following counterexample. Let R^2 be the Euclidean plane, let ϕ_i be a sequence of distinct unit vectors in R^2 which converge to a vector ϕ_0 and let a_i, $i = 0, 1, 2, \ldots$, be the subspaces generated by these vectors. If ϕ is an arbitrary unit vector, m the corresponding state and P_i the projections corresponding to a_i, we have $m(a_i) = \langle \phi, P_i \phi \rangle = \langle \phi, \phi_i \rangle^2 \to \langle \phi, \phi_0 \rangle^2 = m(a_0)$. In this way we see that $m(a_0) = \lim m(a_i)$ for all $m \in M$. However, lim inf $a_i = \{0\} < a_0 <$ lim sup $a_i = R^2$.

We thus see that there are basic differences between the generalized probability theory and the conventional theory. This difference may be illustrated even more strikingly by the following example. In the conventional theory, the states are always subadditive; that is $m(a \vee b) \leq m(a) + m(b)$ for any propositions a, b. This is not necessarily true in the generalized theory. To illustrate this consider the following unit vectors in R^2: $\phi = (1, 0)$, $\phi_1 = (0, 1)$, $\phi_2 = (2)^{-1/2}(1, 1)$. Letting m be the state corresponding to ϕ, and P_1, P_2 the projections on the subspaces a_1, a_2 generated by ϕ_1, ϕ_2 we have $m(a_1 \vee a_2) = 1$ and $m(a_1) + m(a_2) = \langle \phi, P_1 \phi \rangle + \langle \phi, P_2 \phi \rangle = 0 + \frac{1}{2}$ and hence $m(a_1 \vee a_2) > m(a_1) + m(a_2)$!

We next show that Example 2 is a generalization of Example 1. That is, the Hilbert space formulation of generalized probability is more general than the conventional theory.

Theorem 5.4. *If (Ω, F) is a measurable space, there exists a Hilbert space H, and a σ-isomorphism h from F into the lattice L of all closed subspaces of H. For every probability measure μ on (Ω, F) there is a state m_μ on L such that $m_\mu(h(\Lambda)) = \mu(\Lambda)$ for every $\Lambda \in F$. For any class of random variables $\{f_\alpha : \alpha \in A\}$ on (Ω, F) there is a class of observables $\{x_\alpha : \alpha \in A\}$ such that $x_\alpha(E) = h[f_\alpha^{-1}(E)]$ for all $E \in \mathscr{B}(R)$.*

Proof. We shall give a sketch of the proof. Let $\{\mu_\delta : \delta \in D\}$ be the collection of all probability measures on (Ω, F) and let $H_\delta = L_2(\Omega, F, \mu_\delta)$ If $\Lambda \in F$, let $h_\delta(\Lambda) = \text{Cl}\{f \in H_\delta : f(\Lambda') = 0\}$. It is straight forward to show that h_δ is a σ-homomorphism from F into the lattice of all closed subspaces of H_δ. Now $H = \sum \{H_\delta : \delta \in D\}$, the direct sum of the Hilbert spaces H_δ, is a Hilbert space (cf. [7, p. 256]). For $\Lambda \in F$, define $h(\Lambda) = \sum \{h_\delta(\Lambda) : \delta \in D\}$. It is routine to show that h is a σ-homomorphism from F into the lattice L of all closed subspaces of H. To show h is one–one suppose $h(\Lambda) = h(\Gamma)$. Then by definition $h_\delta(\Lambda) = h_\delta(\Gamma)$ for all $\delta \in D$, and it then follows that $\mu_\delta[(\Lambda \cap \Gamma') \cup (\Lambda' \cap \Gamma)] = 0$. But then $(\Lambda \cap \Gamma') = \varnothing$ since if $p \in \Lambda \cap \Gamma'$ and μ is a probability measure concentrated at p, then $\mu(\Lambda \cap \Gamma') = 1$, a contradiction. Similarly $(\Lambda' \cap \Gamma) = \varnothing$ and hence $\Lambda = \Gamma$. Now if μ_β is a probability measure, let $\phi \in H$ be defined by $\phi_\beta \equiv 1$, $\phi_\alpha \equiv 0$, $\alpha \in D - \{\beta\}$. Define the state m_β by $m_\beta(a) = \langle \phi, P_a \phi \rangle$ for all $a \in L$ where P_a is the orthogonal projection onto a. Now it is easily seen that $(P_{h(\Lambda)} \psi)_\delta = \chi_\Lambda \psi_\delta$, $\delta \in D$, for all $\psi \in H$. We therefore have

$$m_\beta[h(\Lambda)] = \langle \phi, P_{h(\Lambda)} \phi \rangle = \sum \{\langle \phi_\delta, (P_{h(\Lambda)} \phi)_\delta \rangle : \delta \in D\}$$

$$= \sum \{\langle \phi_\delta, \chi_\Lambda \phi_\delta \rangle : \delta \in D\} = \langle \phi_\beta, \chi_\Lambda \phi_\beta \rangle$$

$$= \int_\Lambda \langle d\mu_\beta = \mu_\beta(\Lambda).$$

The last statement of the theorem is trivial.

Most of the remainder of this section is concerned with independent observables. A finite collection of observables $x_1, \ldots, x_n$ is *independent in the state* m if

$$m[x_1(E_1) \wedge \cdots \wedge x_n(E_n)] = m[x_1(E_1)]m[x_2(E_2)] \cdots m[x_n(E_n)],$$

for all $E_i \in \mathscr{B}(R)$, $i = 1, 2, \ldots, n$. An arbitrary collection of observables is independent in a state if any finite subset is independent in that state.

Theorem 5.5. *Two observables x and y are independent in every state if and only if x or y is a constant (i.e., has one point spectrum).*

Proof. Sufficiency is trivial; for necessity, let $E, F \in \mathscr{B}(R)$ and let $a = x(E)$, $b = y(F)$. If $a - (a \wedge b) \neq 0$ and $b - (a \wedge b) \neq 0$, then there are states m_1 and m_2 with $m_1[a - (a \wedge b)] = 1$ and $m_2[b - (a \wedge b)] = 1$. Then $m_1(a \wedge b) = m_2(a \wedge b) = 0$ so if $m = \frac{1}{2}m_1 + \frac{1}{2}m_2$ we have $m(a)m(b) > 0$ and $m(a \wedge b) = 0$, which is a contradiction. Thus we have either $x(E) \leq y(F)$ or $y(F) \leq x(E)$. If $0 < x(E) \leq y(F) < 1$, then the inequalities for $x(E)$ and $y(F')$ lead to a contradiction. Thus if $0 < x(E) < 1$, then it follows that for all $F \in \mathscr{B}(R)$ we have $y(F) = 0$ or $y(F) = 1$. It follows that there is a $\lambda \in R$ such that $y(\{\lambda\}) = 1$ and thus y has one point spectrum.

If $x, x_1, x_2, \ldots$ are observables then x_n *converges in measure* (m) to x if $\lim_{n \to \infty} m((x - x_n)([-\varepsilon, \varepsilon]')) = 0$ for every $\varepsilon > 0$; and x_n *converges in mean* (m) to x if $\lim_{n \to \infty} m((x - x_n)^2) = 0$. Recall from Corollary 4.13 that if x_i is a sequence of compatible observables there exists a measure space (Ω, F) a σ-homomorphism h from F into L and F-measurable functions f_i such that $x_i(E) = h(f_i^{-1}(E))$ for every $E \in \mathscr{B}(R)$. We denote this correspondence by $x_i \sim f_i$, and if m is a state on L we define the probability measure $\overline{m}$ on F by $\overline{m}(\Lambda) = m(h(\Lambda))$, $\Lambda \in F$. The proof of the next lemma is straightforward.

Lemma 5.6. *Let x_i be a sequence of simultaneous observables on L and let m be a state.* (i) *The probability distribution $m[x_i(\cdot)]$ of x_i with respect to m is the probability distribution of f_i with respect to $\overline{m}$, where $x_i \sim f_i$.* (ii) *If the x_i's are independent with respect to m, then the f_i are independent with respect to $\overline{m}$.* (iii) *If $x_i \sim f_i$ and $x \sim f$, then $x_i \to x$ in measure (m) if and only if $f_i \to f$ in measure $(\overline{m})$.*

Using this lemma, theorems about random variables apply to one observable or a sequence of compatible ones. Thus, for example, the law of large numbers, the central limit theorem and other limit theorems can be obtained for sequences of independent compatible observables as direct consequences of the corresponding theorems for random variables. We thus see that independence of compatible observables is essentially the same concept as that in the conventional theory. We now show that this is not so in the case of noncompatible observables.

We say that a collection of propositions is *independent* if the corresponding characteristic observables are. Thus $a_1, \ldots, a_n$ are independent in the state m if and only if

$$m(a_1^{(')} \wedge a_2^{(')} \wedge \cdots \wedge a_n^{(')}) = m(a_1^{(')})m(a_2^{(')}) \cdots m(a_n^{(')})$$

where the parentheses indicate that the equality must hold with any of the primes removed, at the same time, from both sides. Thus we have 2^n equalities. The proof of the following lemma is standard.

Lemma 5.7. *If $a \leftrightarrow b$, then a and b are independent in the state m if and only if $m(a \wedge b) = m(a)m(b)$.*

Lemma 5.7 does not hold for noncompatible propositions. To see this let $\phi = (0, 0, 1)$, b be the subspace generated by $(0, 1, 1)$, and a be the subspace generated by $(1, 0, 0)$ and $(0, 1, 0)$. If m is the state corresponding to ϕ, we have $m(a \wedge b) = m(a)m(b) = 0$, but $m(a' \wedge b) = 0$ while $m(a')m(b) \neq 0$.

As an example of noncompatible independent propositions, let $A = \{a_\delta : \delta \in D\}$ be any collection of propositions such that $\bigwedge a_\delta$ exists and $\bigwedge a_\delta \neq 0$. Then A is independent with respect to any state m which satisfies $m(\bigwedge a_\delta) = 1$. It is also easily seen that if $\{A_\delta : \delta \in D\}$ is a set of self-adjoint operators and ϕ is a vector in the null space of all of them then the A_δ's are independent in the state corresponding to ϕ. We can even characterize independent one-dimensional subspaces in a Hilbert space with respect to a pure state.

Theorem 5.8. *Let $A = \{a_\delta : \delta \in D\}$ be distinct one-dimensional subspaces of a Hilbert space H and let m be the state corresponding to a vector $\phi \in H$. Then A is independent with respect to m if and only if A has one of the two forms:*

(i) *$\phi \perp a_\delta$ for all $\delta \in D$,*
(ii) *there is a $\gamma \in D$ such that $\phi \perp a_\delta$, $\delta \in D - \{\gamma\}$ and $a_\gamma \perp a_\delta$, $\delta \in D - \{\gamma\}$.*

Proof. To prove necessity, suppose a, $b \in A$ and $m(a)$, $m(b) \neq 0$. Then $0 = m(a \wedge b) = m(a)m(b)$ which is a contradiction. Therefore $\phi \perp a_\delta$ for all $\delta \in D$ except possibly one. Suppose $m(a_\gamma) \neq 0$, $\gamma \in D$. If a_γ is not orthogonal to a_δ for $\delta \in D - \{\gamma\}$ then $a_\gamma \not\subset a_\delta'$. Hence $0 = m(a_\gamma \wedge a_\gamma') = m(a_\gamma)m(a_\delta') = m(a_\gamma)$ a contradiction. Therefore $a_\gamma \perp a_\gamma$ for all $\delta \in D - \{\gamma\}$. Sufficiency is left to the reader. •

It would be interesting if characterizations of more general independent subspaces could be given. Notice that if $\{a_\delta : \delta \in D\}$ are independent with respect to states m_i, $i = 1, 2, \ldots$, then $\{a_\delta : \delta \in D\}$ need not be independent with respect to the mixture $\sum \lambda_i m_i$, $\lambda_i \geq 0$, $\sum \lambda_i = 1$. This may be seen from the following example. Let a_1, a_2 be subspaces generated by the vectors $\phi_1 = (1, 0)$, $\phi_2 = (0, 1)$, respectively and let m_1, m_2 be the states corresponding to ϕ_1, ϕ_2. Then a_1, a_2 are independent with respect to m_1 and m_2 but not with respect to $\frac{1}{2}m_1 + \frac{1}{2}m_2$.

Cantelli's lemma says that if $\sum m(a_i) < \infty$ then $m(\limsup a_i) = 0$. Although, as is easily seen, this lemma holds if the a_i's are compatible; it does not hold in general. For example, let $\phi = (0, 1)$, $\phi_n = (1, 2^{-n})$, $n = 1, 2, \ldots$; let a_n be the subspace generated by ϕ_n and m the state corresponding to ϕ. Then $m(\limsup a_i) = \lim m(a_k \vee a_{k+1} \vee \cdots) = 1$ and $\sum m(a_i) < \infty$. If the a_i's are independent then not only does Cantelli's lemma hold but also the following stronger result due to Borel and Cantelli.

Theorem 5.9. *If a_i, $i = 1, 2, \ldots$ are independent in the state m, then* $m(\limsup a_i) = 1$ *if* $\sum m(a_i) = \infty$ *and* $m(\limsup a_i) = 0$ *if* $\sum m(a_i) < \infty$.

We have seen that subadditivity holds for compatible propositions but need not hold for noncompatible ones. It is interesting to note that if a and b are independent then subadditivity holds. Indeed,

$$m(a \vee b) = m((a' \wedge b')') = 1 - m(a' \wedge b') = 1 - m(a')m(b')$$

$$= m(a) + m(b) - m(a)m(b) \leq m(a) + m(b).$$

Most of the limit theorems of conventional probability are based on the fact that if x and y are random variables which are independent with respect to a measure μ and F_{x+y}, F_x, F_y are the distribution functions of $x + y$, x, and y, respectively, then

$$F_{x+y}(\lambda) = \int_{-\infty}^{\infty} F_x(\lambda - \xi) \, dF_y(\xi) = (F_x * F_y)(\lambda).$$

Motivated by this, we say that observables $x_1, \ldots, x_n$ are *strongly independent* in the state m if for any n Borel functions $f_1, \ldots, f_n$ we have

$$F(f_1(x_1) + \cdots + f_n(x_n), m; \lambda) = F(f_1(x_1), m; \lambda) * \cdots * F(f_n(x_n), m; \lambda),$$

where $F(z, m; \lambda) = m[z((-\infty, \lambda])]$ is the *distribution function* of an observable z in the state m. As usual, a collection $\{x_\gamma : \delta \in D\}$ of observables is strongly independent in the state m if every finite subcollection is strongly independent in m and a collection of propositions is strongly independent if the corresponding characteristic observables are also.

Theorem 5.10. *If a collection $\{x_\gamma : \delta \in D\}$ of observables is strongly independent in the state m, it is independent in the state m.*

Proof. Let $x_1, \ldots, x_n$ be a finite number of observables in the collection, and let $E(i) \in \mathscr{B}(R)$, $i = 1, 2, \ldots, n$. By an induction argument it can be shown that

$$F(\chi_{E(1)}(x_1) + \cdots + \chi_{E(n)}(x_n), m; \lambda)$$

$$= \begin{cases} 1 - \prod_1^n m[x(E(i))], & n - 1 \leq \lambda < n \\ 1, & \lambda \geq n. \end{cases}$$

Therefore,

$$m([\chi_{E(1)}(x_1) + \cdots + \chi_{E(n)}(x_n)](\{n\})) = \prod_1^n m[x_i(E(i))].$$

Applying Corollary 4.34 (iii) we have

$$[\chi_{E(1)}(x_1) + \cdots + \chi_{E(n)}(x_n)](\{n\}) = x_1(E_1) \wedge \cdots \wedge x_n(E_n).$$

Hence $m[\wedge x_i(E(i))] = \prod_1^n m[x_i(E(i))]$ and thus $x_1, \ldots, x_n$ are independent.

It is clear that a collection of compatible observables is strongly independent if and only if it is independent and thus the two concepts are identical in the conventional theory. However, whether independent observables are strongly independent or not is not known to this author. It seems unlikely that they are. However, if the answer to this question is affirmative, it seems clear that the usual limit theorems concerning sums of independent observables would go through. We can answer the above question affirmatively in the case of one-dimensional subspaces of a Hilbert space since we have a characterization of independence of these propositions.

Theorem 5.11. *A collection of one-dimensional subspaces* $\{a_\delta : \delta \in D\}$ *is strongly independent in a pure state m if and only if it is independent in m.*

Proof. Necessity follows from Theorem 5.10. To prove sufficiency, let $a(1), \ldots, a(n)$ be a finite subcollection of the independent collection $\{a_\delta : \delta \in D\}$ in the pure state m. We have one of the two cases in Theorem 5.8. If the vector ϕ corresponding to m is orthogonal to $a(1), \ldots, a(n)$ then

$$F(x_{a(1)}, m; \lambda) * \cdots * F(x_{a(n)}, m; \lambda) = \begin{cases} 0, & \lambda < 0 \\ 1, & \lambda \geq 0. \end{cases}$$

Now as in Corollary 4.34 (i) $(x_{a(1)} + \cdots + x_{a(n)})(\cdot)$ is a resolution of the identity which takes $\{0\}$ into the projection on the orthogonal complement of $\bigwedge a_i$. Therefore,

$$\begin{aligned} F(x_{a(1)} + \cdots + x_{a(n)}, m; \lambda) &= m[(x_{a(1)} + \cdots + x_{a(n)})((-\infty, \lambda])] \\ &= \langle \phi, (x_{a(1)} + \cdots + x_{a(n)})((-\infty, \lambda])\phi \rangle \\ &= \begin{cases} 0, & \lambda < 0 \\ 1, & \lambda \geq 0. \end{cases} \end{aligned}$$

If we have the second case in Theorem 5.8, reorder the a_i's if necessary so that $a_n \perp a_i$, $1 \leq i < n$ and $\phi \perp a_i$, $1 \leq i < n$. Then

$$F(x_{a(1)}, m; \lambda) * \cdots * F(x_{a(n)}, m; \lambda) = \begin{cases} 0, & \lambda < 0 \\ 1 - m(a_n), & 0 \leq \lambda < 1 \\ 1, & 1 \leq \lambda. \end{cases}$$

As is easily checked, this same equation holds for

$$F(x_{a(1)} + \cdots + x_{a(n)}, m; \lambda).$$

We now turn to a weaker condition than that of strong independence. This condition is quite important in certain applications. Observables $x_1, \ldots, x_n$ are *uncorrelated* in the state m if $m[(x_1 + \cdots + x_n)^2] = \sum_1^n m(x_i^2) + \sum_{i \neq j} m(x_i)m(x_j)$. Notice that for two bounded self-adjoint operators A, B to be uncorrelated in a state corresponding to a vector ϕ it is necessary and sufficient that $\langle \phi, (AB + BA)\phi \rangle = 2\langle \phi, A\phi \rangle \times \langle \phi, B\phi \rangle$. A standard argument shows that strongly independent observables are uncorrelated, and it is well-known from the conventional theory that uncorrelated observables need not be strongly independent. The *variance* of an observable x in the state m is $v_m = m[(x - m(x)I)^2] = m(x^2) - [m(x)]^2$. It follows from a simple computation that if $x_1, \ldots, x_n$ are uncorrelated in m then $v_m(x_1 + \cdots + x_n) = v_m(x_1) + \cdots + v_m(x_n)$.

There is an interesting relation between the point spectrum of an observable and the states in which $v_m = 0$. If $\lambda \in \sigma_p(x)$ we say that m is an *eigenstate* of x corresponding to the *eigenvalue* λ if $m[x(\{\lambda\})] = 1$. Physically, an eigenvalue of an observable x corresponding to the eigenstate m is a value which x attains with certainty in the state m. Intuitively, one would then expect that if m is an eigenstate for x, the uncertainty or variance of x in the state m is zero.

Theorem 5.12. *m is an eigenstate of x if and only if $v_m[u(x)] = 0$ for every Borel function u.*

Proof. Let m correspond to the eigenvalue λ_0. Then $m[u^2(x)] = \int u^2(\lambda) m[x(d\lambda)] = u^2(\lambda_0) = (m[u(x)])^2$. Conversely, we have $v_m[\chi_E(x)] = 0$ for any $E \in \mathscr{B}(R)$ and thus $m[x(E)] = m[\chi_E^2(x)] = (m[\chi_E(x)])^2 = (m[x(E)])^2$. Therefore, $m[x(E)] = 0$ or 1 for every $E \in \mathscr{B}(R)$. It follows that there is a $\lambda_0 \in R$ such that $m[x(\lambda_0)] = 1$ and hence m is an eigenstate.

We close this section with a simple form of the law of large numbers for uncorrelated observables.

Theorem 5.13. *If $x_1, x_2, \ldots$, are uncorrelated in the state m and if there is a constant K such that $m(x_i^2) \le K$, $i = 1, 2, \ldots$, then*

$$(x_1 + \cdots + x_n)/n \to m[(x_1 + \cdots + x_n)/n]I$$

in measure and in mean.

Proof. By Schwartz's inequality we have

$$|m(x_i)| \le \int |\lambda| \, m[x_i(d\lambda)] \le \left(\int \lambda^2 m[x_i(d\lambda)] \right)^{1/2} = [m(x_i^2)]^{1/2}.$$

We then obtain

$$m[((x_1 + \cdots + x_n)/n - m((x_1 + \cdots + x_n)/n)I)^2]$$

$$= v_m[(x_1 + \cdots + x_n)/n] = n^{-2} \sum_1^n v_m(x_i)$$

$$= n^{-2} \left[\sum_1^n m(x_i^2) - \sum_1^n m(x_i)^2 \right]$$

$$\le 2n^{-2} \sum_1^n m(x_i^2) \le 2K/n.$$

Therefore we have convergence in mean (m) and hence also convergence in measure (m). The material presented here may be found in [15], where the reader is referred for further details.

One should note that the concepts of convergence almost everywhere, everywhere, almost uniformly, uniformly, etc. may be extended to the generalized theory. For example, we say that a sequence of observables x_i converges *everywhere* to an observable x if lim sup$\{(x_i - x)([-\varepsilon, \varepsilon]')\}$ $= 0$ for every $\varepsilon > 0$, and x_i converges *almost everywhere* (m) to x if $m(\text{lim sup}\{(x_i - x)([-\varepsilon, \varepsilon]')\}) = 0$ for every $\varepsilon > 0$. We say that $x_i \to x$ *uniformly* if for any $\varepsilon > 0$ there is an n such that $i \geq n$ implies $(x_i - x)$ $([-\varepsilon, \varepsilon]') = 0$; and $x_i \to x$ *almost uniformly* (m) if for any $\delta > 0$ there is a proposition a such that $m(a') < \delta$ and such that for any $\varepsilon > 0$ there is an n for which $i \geq n$ implies $(x_i - x)([-\varepsilon, \varepsilon]) \geq a$. These definitions reduce to the usual ones if the logic is Example 1. Types of convergences similar to these have been defined and studied by Segal and others [48, 53] for a different noncommutative measure theory than that considered here.

It is interesting to compare the properties of these types of generalized convergences to their classical counterparts. For example, it is clear that uniform convergence implies all other kinds of convergence and is equivalent to convergence under the spectral norm. Also one easily sees that almost uniform (m) convergence implies almost everywhere (m) convergence which implies convergence in measure (m). Unlike the classical theory it is not true that if a sequence converges in measure there exists a subsequence which converges almost everywhere. (The details of these results and others may be found in [39a].) One can prove the following bounded convergence theorem. If $x_n \to x$ in measure (m) and there exists a $K < \infty$ such that $|x_n| < K$ for every n, then $m(x_n) \to m(x)$ as $n \to \infty$. Also a kind of Fatou's lemma may be proved. Surprisingly enough, it can be shown that Egoroff's theorem does not hold.

In Section IV.B we introduced the notion of a joint distribution of two observables. This notion can be weakened to a considerable extent and there is quite a bit of literature on the subject. For example, Wigner [60a] and Moyal [39a] proceed as follows. In ordinary probability theory, knowing the distribution μ_f of a random variable f is equivalent to knowing the characteristic function

$$\phi_f(t) = \int_{-\infty}^{\infty} e^{it\lambda} \mu_f(d\lambda) = \text{Av}(e^{itf})$$

of f. In fact, we can obtain μ_f from ϕ_f using the inversion formula

$$\mu_f([a, b]) = \lim_{U \to \infty} \frac{1}{2\pi} \int_{-U}^{U} \frac{e^{-iua} - e^{-iub}}{iu} \phi_f(u) \, du.$$

If μ_f has a density function h we get $\phi_f(t) = \int_{-\infty}^{\infty} e^{it\lambda} h(\lambda) \, d\lambda$ and therefor by Fourier inversion we have

$$h(\lambda) = \frac{1}{2\pi} \int_{-\infty}^{\infty} \phi_f(t) e^{-it\lambda} \, dt.$$

Similarly if f and g are random variables the joint characteristic function is

$$\phi(s, t) = \int_{-\infty}^{\infty} \int (\exp[isu + itv]) \mu_{f,g}(du \, dv) = \text{Av}(\exp[isf + itg])$$

where $\mu_{f,g}$ is the joint distribution of f and g, and there is a corresponding two-dimensional inversion formula. If $\mu_{f,g}$ has a density function $h(u, v)$ then

$$\phi(s, t) = \int_{-\infty}^{\infty} \int (\exp[isu + itv]) h(s, t) \, ds \, dt$$

and

$$h(s, t) = \frac{1}{4\pi^2} \int_{-\infty}^{\infty} \int (\exp[-isu - itv]) \phi(u, v) \, du \, dv.$$

Returning to quantum mechanics, let us suppose that A and B are self-adjoint operators representing observables and that ψ is a unit vector representing a state. Ignoring questions concerning the domains of the operators we define the joint characteristic function of A, B in the state ψ to be $\phi(s, t) = \text{Av}(\exp[isA + itB]) = \langle \psi, (\exp[isA + itB]) \psi \rangle$. We define the density function as

$$h(s, t) = \frac{1}{4\pi^2} \int_{-\infty}^{\infty} \int (\exp[-isu - itv]) \phi(u, v) \, du \, dv$$

$$= \frac{1}{4\pi^2} \int_{-\infty}^{\infty} \int \exp[-isu - itv] \langle \psi, (\exp[iuA + ivB]) \psi \rangle \, du \, dv.$$

Thus the *weak joint distribution* (in the sense of Wigner) becomes $\mu_{A,B}(E) = \int_E \int h(s, t) \, ds \, dt$ for all $E \in \mathscr{B}(R^2)$. Wigner has shown that for a coordinate observable Q and momentum observable P which are known to satisfy $QP - PQ = ihI$ we have, for the state $\psi \in L_2(R)$,

$$h_{Q, P}(s, t) = \frac{1}{2\pi} \int_{-\infty}^{\infty} \psi^*(s - \tfrac{1}{2}\hbar u) \psi(s + \tfrac{1}{2}\hbar u) e^{-iut} \, du.$$

For most ψ's one can see that $h_{Q,P}$ can have negative values and hence so does $\mu_{Q,P}$. This makes $\mu_{Q,P}$ a highly unsatisfying joint distribution.

Another approach due to Urbanik [56a] and Varadarajan [57, 57a] goes as follows. It is well-known that if f and g are random variables on (Ω, F, μ) then the distributions of f and g do not determine their joint distribution. However, the distributions of $rf + sg$ for every $r, s \in R$ do determine the joint distribution. In fact, the joint distribution $\mu_{f,g}$ is the unique measure on $\mathscr{B}(R^2)$ that satisfies $\mu_{f,g}\{(\omega_1, \omega_2): r\omega_1 + s\omega_2 \in E\} = \mu\{\omega \in \Omega: rf(\omega) + sg(\omega) \in E\}$ for every $E \in \mathscr{B}(R)$. Now suppose A and B are self-adjoint operators and ψ a unit vector. Then the *weak joint distribution* (in the sense of Urbanik) of A and B in the state ψ is said to exist if there is a two-dimensional Borel measure $\mu_{A,B}$ such that $\mu_{A,B}\{(\omega_1, \omega_2): r\omega_1 + s\omega_2 \in E\} = \langle \psi, P^{rA+sB}(E)\psi \rangle$ for every $E \in \mathscr{B}(R)$ where $P^{rA+sB}(\cdot)$ is the resolution of identity of $rA + sB$. Of course, for this to be rigorous we would have to require certain properties of the common domain of A and B so that $rA + sB$ is self-adjoint for all $r, s \in R$. Varadarajan has shown that if A and B have a common dense domain then $\mu_{A,B}$ exists for every state in the common domain if and only if A and B commute.

We now give still another approach [17a] which seems more natural to the author. This approach is easily generalized to abstract observables, works for any two self-adjoint operators and does not depend upon their domains. Recall that the joint distribution of two random variables f and g on (Ω, F, μ) is the unique measure $\mu_{f,g}$ on $\mathscr{B}(R^2)$ such that $\mu_{f,g}(E \times F) = \mu(f^{-1}(E) \cap g^{-1}(F))$, $E, F \in \mathscr{B}(R)$. If A and B are self-adjoint operators we say that A and B have a *weak joint distribution in the state* ψ if there is a probability measure $\mu_{A,B}$ on $\mathscr{B}(R^2)$ such that $\mu_{A,B}(E \times F) = \langle \psi, P^A(E) \wedge P^B(F)\psi \rangle$ for every $E, F \in \mathscr{B}(R)$. Even though two random variables always have a joint distribution we can show that two observables need not have a weak joint distribution in this sense. In fact, we can give a necessary and sufficient condition on a state ψ so that A and B have a weak joint distribution in this state.

Theorem 5.14. *Two self-ajdoint operators A, B have a weak joint distribution in the state ψ if and only if $P^A(E)P^B(F)\psi = P^B(F)P^A(E)\psi$ for every $E, F \in \mathscr{B}(R)$.*

For the proof of this theorem and generalizations to abstract observables the reader is referred to [17a].

VI. An Axiomatic Model for Quantum Mechanics

A. THE GENERAL AXIOMS

Thus far we have considered only the logic of quantum mechanics, a generalized probability theory and some consequences of the abstract definitions. We now collect axioms to form a model for quantum mechanics. Let S be the *physical space* corresponding to a laboratory experiment. For example in the case of a system with a finite number of degrees of freedom, S would be the configuration space of our system. Mathematically we shall define physical space S to be a locally compact Hausdorff space with second countability. Let G be the *group of rigid motions* on S. For example in the finite degrees of freedom case G would be the group generated by the translations, rotations, and reflections in a finite dimensional Euclidean space. Mathematically we define the group of rigid motions G to be a locally compact group with second countability which forms a continuous, effective, transitive transformation group on S. That is, there exists a map from $S \times G$ onto S denoted by $(s, g) \to sg$, $g \in G$, $s \in S$ such that

(G-1) if $s_1, s_2 \in S$ there is a $g \in G$ such that $s_1 = s_2 g$ (transitivity);
(G-2) for every $g \in G$, $s \to sg$ is a homeomorphism of S with itself;
(G-3) $(sg_1)g_2 = (s)g_1 g_2$ for every $g_1, g_2 \in G$, $s \in S$;
(G-4) $(s)g = s$ for every $s \in G$ if and only if $g = e$ where e is the identity of G (effectiveness).

Note that our finite degrees of freedom example satisfies these conditions. Before we postulate our axiomatic model we need one more definition. A *unitary derivable* is a σ-homomorphism from the Borel sets of the complex plane $\mathscr{B}(C)$ into L whose spectrum is contained in the unit circle $\{\lambda : |\lambda| = 1\}$. The theory of (unitary) derivables is the same as that of observables and all the theorems we have proved for observables go through for derivables with the obvious modifications. We now state three postulates.

1. The propositions and states of a quantum mechanical system form a logic (L, M) satisfying condition C.
2. There is a σ-homomorphism X from the Borel sets $\mathscr{B}(S)$ of physical space S into L such that $g \to m[X(\Lambda g)]$ is continuous for every $m \in M$, $\Lambda \in \mathscr{B}(S)$.
3. For every g in the group of rigid motions G there is a unitary

derivable x_g such that
   (i) $g \rightarrow m(x_g)$ is continuous for every $m \in M$;
   (ii) if $gg' = g'g$ then $x_g \leftrightarrow x_{g'}$ and $x_{gg'} = x_g x_{g'}$;
   (iii) $X(\Lambda) = X(\Lambda g)$ if and only if $X(\Lambda) \leftrightarrow x_g(E)$ for every $E \in \mathscr{B}(C)$.

Notice that $m(x_g)$ is the expectation of x_g defined in the usual way and $x_g x_{g'}$ is the product of compatible derivables defined analogously to the product of compatible observables. The physical justification of Postulate 1 has already been discussed in Section III. Postulate 2 is necessary to ensure the existence of coordinate observables. The σ-homomorphism $X: \mathscr{B}(S) \rightarrow L$ is called the *position σ-homomorphism* and $X(\Lambda)$ may be thought of as the proposition that the position of the system is in the set $\Lambda \subset S$. In this way Postulate 2 gives a way of representing the position of the system in the logic. Postulate 3 is necessary to ensure the existence of momentum observables and gives a way of exhibiting the group action in the quantum mechanical logic. Postulate 3(ii) says that if g and g' commute, then their actions on S can be observed at the same time. That is, the action described by g does not affect the action described by g' and vice versa. The rest of Postulate 3(ii) states that the action described by the product of g and g' is the product of the actions. Postulate 3 (iii) just gives two ways of saying that the proposition $X(\Lambda)$ is unaffected by the action of g on S. The reader may wonder why x_g is a unitary derivable and not, as would seem more natural, an observable. The fact is that our theory can be carried through under the assumption that x_g is an observable. However, we assume x_g is a unitary derivable because we want our theory to include the usual quantum mechanical formulation in which it is usually assumed that G has a continuous unitary representation on the Hilbert space of states. If $u: S \rightarrow R$ is a Borel function and X a position σ-homomorphism we define $u(X)$ as might be expected, by $u(X)(E) = X(u^{-1}(E))$, $E \in \mathscr{B}(R)$. Thus $u(X)$ is an observable.

The reader should note that Axioms 1, 2, 3(i), and 3(ii) apply to classical systems as well as quantum systems. Thus as far as these axioms are concerned L might be a Boolean σ-algebra. However, Axiom 3(iii) ushers us into a purely quantum mechanical realm. Specifically, we use the fact that $X(\Lambda) \leftrightarrow x_g(E)$ for all $E \in \mathscr{B}(C)$ to describe the situation that the propositions $X(\Lambda)$ and $X(\Lambda g)$ are identical. Thus for the case in which $X(\Lambda) \neq X(\Lambda g)$ which is, of course, quite common we must have $X(\Lambda) \leftrightarrow x_g(E)$ for some $E \in \mathscr{B}(C)$, and the latter is a purely quantum mechanical phenomenon. To emphasize this fact we give the following theorem which shows that L cannot be a Boolean σ-algebra unless the space S is trivial. To avoid pathologies we always assume $0 \neq 1$ in L (i.e., L has at least two elements).

Theorem 6.1. *Suppose we have a system satisfying Axioms 1–3. If L is a Boolean σ-algebra, then S contains only one point.*

For the proof of the above theorem and other details which are omitted from this section the reader is referred to [16]. We shall see later that another important consequence of 3(iii) is that the conjugate coordinate and momentum observables are noncompatible.

A *coordinate system* for physical space S is a collection of real valued continuous functions $\{f_\alpha : \alpha \in A\}$ on S such that $f_\alpha(s_1) = f_\alpha(s_2)$ for all $\alpha \in A$ implies $s_1 = s_2$. For simplicity, the coordinate systems that we are considering here are generalizations of rectangular coordinate systems. If $\{f_\alpha : \alpha \in A\}$ is a coordinate system, a subset G_β, $\beta \in A$, of the group of rigid motions G is a *motion in the β direction* if $f_\alpha(sg) = f_\alpha(s)$, for all $s \in S$, $g \in G_\beta$, and $\beta \neq \alpha \in A$; and $f_\beta(sg_1 g_2) = f_\beta(sg_1) + f_\beta(sg_2) - f_\beta(s)$, for all $s \in S$, g_1, $g_2 \in G_\beta$. $G_\beta \subset G$ is a *group motion* in the β direction if G_β is a closed Abelian group which is a motion in the β direction. It is shown in [16] that any motion in the β direction is contained in a group motion in the β direction. Now a group motion G_β is a subgroup of G which corresponds physically to a movement of the system in the β direction and must therefore be associated in some sense with the momentum in the β direction. Since there may be many group motions in the β direction and since the momentum should be uniquely defined, we shall look for some kind of invariant of the group motion. It will turn out that such an invariant is a one-parameter subgroup of the group motion. One is also led to one-parameter subgroups from analogy with classical mechanics (cf. Section III). If $G_1 \subset G$ we say that s_1, $s_2 \in S$ are *connected* by G_1, if there is a $g \in G_1$ such that $s_1 = s_2 g$. A motion in the β direction P_β is a *one-parameter motion in the β direction* if there is a continuous map $\lambda \to g_\lambda$ from $(-\infty, \infty)$ onto P_β such that $g_{\lambda + \mu} = g_\lambda g_\mu$ and if for any s_1, $s_2 \in S$ which are not connected by P_β we have

$$f_\beta(s_1 g) - f_\beta(s_1) = f_\beta(s_2 g) - f_\beta(s_2), \qquad \text{for all} \qquad g \in P_\beta. \quad (6.1)$$

We now show that Eq. (6.1) automatically holds if s_1, s_2 are connected by P_β. Indeed, in this case there is a $g_0 \in P_\beta$ such that $s_1 = s_2 g_0$. Hence $f_\beta(s_1 g) - f_\beta(s_1) = f_\beta(s_2 g_0 g) - f_\beta(s_2 g_0) = f_\beta(s_2 g) - f_\beta(s_2)$ for all $g \in P_\beta$. The next theorem characterizes one-parameter motions.

Theorem 6.2. *A continuous map* $\lambda \to g_\lambda$ *from* $(-\infty, \infty)$ *into G is a one-parameter motion in the β direction if and only if there is a real number c such that* $f_\beta(sg_\lambda) = c\lambda + f_\beta(s)$ *and* $f_\alpha(sg_\lambda) = f_\alpha(s)$, $\alpha \neq \beta$, *for every* $s \in S$.

Proof. To prove necessity fix $s \in S$ and define the function $h: R \to R$ by $h(\lambda) = f_\beta(sg_\lambda)$. Then $h(\lambda + \mu) = f_\beta(sg_{\lambda+\mu}) = f_\beta(sg_\lambda g_\mu) = h(\lambda) + ((\mu) - f_\beta(s)$. Since h is continuous it must have the form $h(\lambda) = c(s)\lambda + f_\beta(s)$ for some $c(s) \in R$. Hence $f_\beta(sg_\lambda) = c(s)\lambda + f_\beta(s)$. It follows from Eq. (6.1) that $c(s)$ is independent of s and thus the necessity is proved. To prove sufficiency we have $f_\alpha(sg_{\lambda+\mu}) = f_\alpha(s) = f_\alpha(sg_\lambda g_\mu)$ for $\alpha \neq \beta$. Also

$$f_\beta(sg_\lambda g_\mu) = f_\beta((sg_\lambda)g_\mu) = c\mu + f_\beta(sg_\lambda) = c(\lambda + \mu) + f_\beta(sg) = f_\beta(s_{\lambda+\mu}).$$

Therefore $(s)g_{\lambda+\mu} = (s)g_\lambda g_\mu$ for all $s \in S$ and by effectiveness (G-4), $g_{\lambda+\mu} = g_\lambda g_\mu$. To show we have a motion, $f_\beta(sg_\lambda g_\mu) = f_\beta(sg_{\lambda+\mu}) = c(\lambda + \mu) + f_\beta(s) = f_\beta(sg_\lambda) + f_\beta(sg_\mu)f_\beta - (s)$ and also Eq. (6.1) holds.

Corollary 6.3. *Let* $\{f_\alpha : \alpha \in A\}$ *be a coordinate system,* P_β *a one-parameter motion in the β direction and X a position σ-homomorphism. Then there is a real number c such that* $X[f_\beta^{-1}(E)g_\lambda] = f_\beta(X)(E + c\lambda)$ *for every* $g_\lambda \in P_\beta$, $E \in \mathcal{B}(R)$.

If $G_0 = \{g_\alpha : \alpha \in A\}$ is the motion $g_\alpha = e$, for all $\alpha \in A$, then G_0 is called the *trivial motion*. It is easily seen that a one-parameter motion is the trivial motion if and only if $c = 0$. Two one-parameter motions $P_1 = \{g_\lambda{}^1\}$, $P_2 = \{g_\lambda{}^2\}$ are *equivalent* (written $P_1 \sim P_2$) if there is a nonzero real number μ such that $g_\lambda{}^2 = g_{\mu\lambda}^1$ for all $\lambda \in (-\infty, \infty)$. We easily see that $\sim$ is an equivalence relation. Therefore the set of one-parameter motions is partitioned into equivalence classes, with each one-parameter motion contained in one and only one class. Physically two equivalent one-parameter motions are essentially the same since one results from the other by a linear change of scale.

Corollary 6.4. *Two nontrivial one-parameter motions are equivalent if and only if they are in the same direction.*

We thus see that the nontrivial one-parameter motions are invariants which depend only upon the direction. If a group motion admits a nontrivial one-parameter subgroup then we call it a *conjugate group motion*. If a coordinate function f_β admits a conjugate group motion in the β direction it is called a *conjugate coordinate function*. Not every

group motion is conjugate. Obviously the trivial motion is not conjugate
Now if $\{f_\alpha: \alpha \in A\}$ is a coordinate system and X the position σ-homo-
morphism, the *coordinate observables* are the observables $f_\alpha(X)$, $\alpha \in A$.
Of course, all coordinate observables are compatible. Now suppose f_β
is a conjugate coordinate function and P_β the essentially unique nontrivial one-parameter motion corresponding to f_β. Applying Postulate
3 there are unitary derivables $x_\lambda = x_{g(\lambda)}$, $\lambda \in (-\infty, \infty)$ which satisfy
$x_{\lambda+\mu} = x_{g(\lambda+\mu)} = x_{g(\lambda)} x_{g(\mu)} = x_\lambda x_\mu$ and $\lambda \to m(x_\lambda)$ is continuous. Thus
$\{x_\lambda: \lambda \in (-\infty, \infty)\}$ is a continuous one-parameter group of unitary
derivables and hence x_λ has an infinitesimal generator (see [16]). That
is, there exists an observable p_β such that $x_\lambda = \exp[i\lambda p_\beta]$, $\lambda \in (-\infty, \infty)$.
p_β is called the *momentum observable in the β direction*. Strictly speaking,
for physical reasons, we should define $\hbar p_\beta$ as the momentum observable;
however, we suppress the $\hbar$ here for simplicity. $f_\beta(X)$ and p_β are called
conjugate coordinate and momentum observables.

Theorem 6.5. *If $f_\alpha(X)$ and p_α, $\alpha \in A$, are conjugate coordinate and
momentum observables, then $f_\beta(X) \leftrightarrow p_\alpha$, $\beta \neq \alpha$, but $f_\alpha(X) \nleftrightarrow p_\alpha$.*

Proof. Let g_λ be a nontrivial one-parameter motion in the α direction,
and let x_λ be the corresponding unitary derivables, $\lambda \in (-\infty, \infty)$.
Suppose $\beta \neq \alpha$ and $E \in \mathcal{B}(R)$. If $s \in f_\beta^{-1}(E)$, then $f_\beta(sg_\lambda^{-1}) = f_\beta(s) \in E$
and $sg_\lambda^{-1} \in f_\beta^{-1}(E)$. Hence $s \in f_\beta^{-1}(E)g_\lambda$ and thus $f_\beta^{-1}(E) \subset f_\beta^{-1}(E)g_\lambda$ for
all $\lambda \in (-\infty, \infty)$. The inclusion in the other direction is trivial so
$f_\beta^{-1}(E) = f_\beta^{-1}(E)g_\lambda$ for all $\lambda \in (-\infty, \infty)$. By Postulate 3(iii) we have
$f_\beta(X) \leftrightarrow x_\lambda$, $\lambda \in (-\infty, \infty)$. It now follows (see [16]) that $f_\beta(X) \leftrightarrow p_\alpha$.
Let c be the constant corresponding to g_λ in Theorem 6.2 and suppose
that $f_\alpha(X) \leftrightarrow x_\lambda$ for all $\lambda \in (-\infty, \infty)$. By Postulate 3(iii) and Corollary
6.3 $f_\alpha(X)(E) = X[f_\alpha^{-1}(E)g_\lambda] = f_\alpha(X)(E + c\lambda)$ for all $\lambda \in (-\infty, \infty)$,
$E \in \mathcal{B}(R)$. Now $f_\alpha(X)((-\infty, 0]) = 1$ since if $f_\alpha(X)((-\infty, 0]) < 1$ then
$f_\alpha(X(R)) < 1$, which is impossible. Similarly $f_\alpha(X)((-\infty, -n]) = 1$ for
$n = 1, 2, \ldots$. Therefore

$$1 = \bigwedge \{f_\alpha(X)((-\infty, -n]): n = 1, 2, \ldots\} = f_\alpha(X)\left(\bigcap_{n=1}^{\infty}(-\infty, -n]\right)$$
$$= f_\alpha(X)(\varnothing) = 0$$

which is a contradiction. Therefore $f_\alpha(X) \nleftrightarrow x_\lambda$ for some $\lambda \in (-\infty, \infty)$
and it now follows that $f_\alpha(X) \nleftrightarrow p_\alpha$.

A σ-finite measure ν on $\mathcal{B}(S)$ is *quasi-invariant* with respect to G if
$\nu(\Lambda) = 0$ implies $\nu(\Lambda g) = 0$ for every $g \in G$, $\Lambda \in \mathcal{B}(S)$. It is shown in
[30, 31] that nontrivial quasi-invariant measures always exist and that

if v and v_1 are quasi-invariant measures, they have the same null sets and also $L_2(S, v)$ and $L_2(S, v_1)$ are unitarily equivalent. Let v be a quasi-invariant measure on $\mathscr{B}(S)$ which shall remain fixed in this discussion. It is shown in [16] that there exists a map $m \to \hat{m}$ of M into $L_2(S, v)$ such that $m[X(\Lambda)] = \int_\Lambda (\hat{m})^2 \, dv$. Now $\hat{m}$ is a kind of position probability density and hence $\int \hat{m}(s)\hat{m}(sg) \, dv(s)$ may be thought of as giving the probability that the system has moved from $s \in S$ to sg. Intuitively one might expect $m(x_g)$ to be this same probability. We say that a state m is *canonical* with respect to the measure v if $m(x_g) = \int \hat{m}(s)\hat{m}(sg) \, dv(s)$ for every $g \in G$.

In the examples of quantum mechanical systems known to the author there are always an abundance of canonical states. For example, let S be n-dimensional Euclidean space R^n, L the lattice of orthogonal projections on $L_2(S, v)$ where v is n-dimensional Lebesgue measure, and G the group generated by the translations, rotations and reflections on R^n. For $\Lambda \in \mathscr{B}(S)$, define $X(\Lambda)$ to be multiplication by the characteristic function χ_Λ, and let M be the states on L defined in the usual way. Then S, G, L, M satisfy Axioms 1 and 2. For $g \in G$, let U_g be the operator on $L_2(S, v)$ defined by $U_g\psi(s) = \psi(sg)$. Then U_g is a unitary operator and let the resolution of identity for U_g be denoted by x_g. With the map $g \to x_g$ our system satisfies Axioms 1–3. Now let m be a state corresponding to a unit vector ψ_m. Then one can show [16] that $\hat{\psi}_m(s) = |\psi_m(s)|$. Thus in order for m to be canonical we must have $\int \psi_m(s)\overline{\psi}_m(sg) \, dv(s) = \int |\psi_m(s)| \, |\psi_m(sg)| \, dv(s)$ for all $g \in G$. Hence if $\psi_m(s) \geq 0$, $s \in S$, then ψ_m is canonical. We thus see that there is an abundance of canonical states in this example. In fact $L_2(S, v)$ is the linear hull of vectors corresponding to canonical states.

The next theorem shows that the statistical properties of a quantum mechanical system, in a canonical state, satisfying Axioms 1–3 are described by operators in a Hilbert space.

Theorem 6.6. *Let (L, M) be a logic and let $f_\alpha(X)$, p_α be conjugate coordinate and momentum observables. Then there is a Hilbert space H, a map $m \to \hat{m}$ from M into H which preserves convex combinations in the sense that $[(\sum \lambda_i m_i)^\wedge]^2 = \sum \lambda_i(\hat{m}_i)^2$ and self-adjoint operators S_α, T_α on H such that $m(f_\alpha(X)) = \langle \hat{m}, S_\alpha \hat{m} \rangle$ and if m is canonical $m(p_\alpha) = \langle \hat{m}, T_\alpha \hat{m} \rangle$ when these expressions exist.*

Up to this point we have been discussing quantum statics; that is, the relationship between states and observables at a particular instant of time. It is obviously important to consider quantum dynamics, the way in which states change with the passage of time. Despite the acausal character of quantum mechanics, reflected in the impossibility of making more than statistical statements about the results of measurements of observables, the relation between states at different times is assumed to be causal. Thus the state at time $t_2 \geq t_1$ is uniquely determined by $t_2 - t_1$ and the state at t_1. Therefore, just as in classical mechanics (cf. Section II) there is a one-parameter semigroup of transformations $v_t \colon M \to M$, and as in classical mechanics we shall assume our system is reversible; that is, each v_t is one–one onto. Defining $v_{-t} = (v_t)^{-1}$, $t \to v_t$ becomes a one-parameter group. Let $m_i \in M$, and $\lambda_i \geq 0$, $\sum \lambda_i = 1$. Suppose at $t = 0$ we are in the state m_j with probability λ_j. Then at $t = t_0$ we are in the state $v_{t_0}(m_j)$ with probability λ_j. Thus if we are in the state $\sum \lambda_i m_i$ at $t = 0$ we are in the state $\sum \lambda_i v_{t_0}(m_i)$ at $t = t_0$, and we have $v_t(\sum \lambda_i m_i) = \sum \lambda_i v_t(m_i)$. Hence v_t preserves mixtures of states. It is also physically reasonable that $v_t(m)(a)$ should change by a small amount after small time intervals. We thus define a *dynamical group* on (L, M) as a one-parameter group $v_t \colon M \to M$ which preserves mixtures and for which $t \to v_t(m)(a)$ is continuous for all $m \in M$, $a \in L$.

B. THE HILBERT SPACE AXIOM

Until now our axioms have been relatively intuitive and reasonable from a physical point of view. We now invoke an axiom which is to a certain extent *ad hoc* but which we need to give us a stronger structure in our quantum logic. If L_1 and L_2 are logics, a *logic isomorphism h* is a one–one map of L_1 onto L_2 satisfying (i) $h(1) = 1$, (ii) if $a \perp b$ then $h(a) \perp h(b)$, (iii) $h(\sum a_i) = \sum h(a_i)$. If $L_1 = L_2$ then h is called a *logic automorphism*. We denote the automorphisms on L by auto(L). We have seen in the last subsection that, so far as the statistical properties of coordinate and momentum observables are concerned, in canonical states, that we may consider L to be the lattice of closed subspaces of a Hilbert space. We now make the stronger postulate that we have an isomorphism.

(L-10)   The logic of a quantum mechanical system is isomorphic to the logic of all closed subspaces of a separable, infinite dimensional complex Hilbert space.

Although (L-10) has not been justified physically one can show mathematically that a logic is close to satisfying (L-10) already and that there are not too many other logics which are mathematically elegant and have nice physical properties. If one imposes some additional conditions on a logic L satisfying (L-1)–(L-9) then results close to (L-10) hold. For example, if L is a modular lattice $(a \wedge (b \vee c) = (a \wedge b) \vee c$ if $c \leq a)$ and if all chains $a_1 < a_2 < \cdots < a_n$ have finite length, then L is isomorphic to a direct sum of lattices of the form $L_{F,n}$ and certain lattices in which no chain has length greater than 4, where $L_{F,n}$ is the lattice of subspaces of an n-dimensional vector space over a field F. This last result is due to von Neumann. Piron [42] has shown that if L is a complete atomic lattice it is isomorphic to a direct integral $\int L_\rho$ of lattices of closed subspaces of Hilbert spaces H_ρ over a field. Now for physical reasons we can restrict the possible fields. If the field is continuous and connected to the unit in a certain sense then we are left with the reals, complexes, or quaternions. The work of Stueckleberg and Emch [10] has shown that the other two are "essentially" equivalent to the complexes. Thus our only physical assumption is that there is only one summand in the direct integral. This amounts physically to assuming that every nonconstant observable fails to be compatible with some observable. We have already called such systems coherent (Section IV.E). If there is more than one summand physicists say there are "superselection rules" [19, 21]. Wicks *et al.* [60] have shown that there are quantum systems which have superselection rules; however, we shall usually consider the simpler coherent case.

Due to (L-10) we may now think of observables as corresponding to self-adjoint operators and states as corresponding to unit vectors or mixtures of unit vectors. If A, B are self-adjoint operators then the corresponding observables x, y are their resolutions of the identity. Notice that we usually do not distinguish between a closed subspace and the unique orthogonal projection onto it. We say that A and B *commute* if $x(E)y(F) = y(F)x(E)$ for all $E, F \in \mathcal{B}(R)$.

Lemma 6.7. *Two observables x and y are compatible if and only if their corresponding self-adjoint operators A, B commute.*

Proof. If $x \leftrightarrow y$ and $E, F \in \mathcal{B}(R)$ then there are mutually orthogonal projections P_1, P_2, P_3 such that $x(E) = P_1 + P_2$, $y(F) = P_2 + P_3$. Therefore, $x(E)y(F) = (P_1 + P_2)(P_2 + P_3) = P_3 = y(F)x(E)$ and hence A and B commute. Conversely, if A and B commute, and $E, F \in \mathcal{B}(R)$, then $x(E)y(F)$ is a projection. Thus $x(E) = [x(E) - x(E)y(F)] + x(E)y(F)$, $y(F) = [y(F) - x(E)y(F)] + x(E)y(F)$ and $x \leftrightarrow y$.

We now give a Hilbert space derivation of the Heisenberg uncertainty principle.

Theorem 6.8. *Let x and y be observables with associated self-adjoint operators A, B and let m be a state corresponding to a vector ϕ. Then the product of the variances satisfy $v_m(x)v_m(y) \geq \frac{1}{2}|\langle\phi, (AB - BA)\phi\rangle|$ if all of these quantities exist.*

Proof. Suppose $m(x) = \langle\phi, A\phi\rangle = m(y) = \langle\phi, B\phi\rangle = 0$. Then

$$|\langle\phi, (AB - BA)\phi\rangle| = |\langle A\phi, B\phi\rangle - \langle B\phi, A\phi\rangle|$$
$$= 2\,|\mathrm{Im}\langle A\phi, B\phi\rangle| \leq 2\|A\phi\|\,\|B\phi\|$$
$$= 2\langle\phi, A^2\phi\rangle^{1/2}\langle\phi, B^2\phi\rangle^{1/2} = 2v_m(x)v_m(y).$$

If $m(x), m(y) \neq 0$, then replace A by $A - \langle\phi, A\phi\rangle$ and B by $B - \langle\phi, B\phi\rangle$ to get the same result.

The above theorem even holds for mixed states. Thus we see that $AB - BA$ gives a measure of how compatible the corresponding observables are. As we shall see if x, y are coordinate and momentum observables in the same direction then the corresponding self-adjoint operators A, B satisfy $AB - BA = i\hbar I$ when the left-hand-side is defined. So from the above theorem $v_m(x)v_m(y) \geq \hbar/2$ which is the usual form of the uncertainty principle.

We now give a more elegant description of mixed states in terms of operators. Let A be a bounded self-adjoint operator with pure point spectrum. For such operators there is a complete orthonormal set $\phi_1, \phi_2, \ldots,$ of eigenvectors and suppose $\lambda_1, \lambda_2, \ldots,$ is the corresponding set of eigenvalues. If $\sum|\lambda_i| < \infty$ we say that A is a *trace class* operator and set $\mathrm{tr}(A) = \sum\lambda_i$. More generally let C be any bounded linear operator and let $A = \frac{1}{2}(C + C^*)$, $B = (1/2i)(C - C^*)$. Then A and B are bounded self-adjoint operators and $C = A + iB$. We say that C is a *trace class* operator if both A and B are and set $\mathrm{tr}(C) = \mathrm{tr}(A) + i\,\mathrm{tr}(B)$. It is not hard to show that whenever C is a trace class operator and $\phi_1, \phi_2, \ldots,$ is a complete orthonormal set then $\mathrm{tr}(C) = \sum\langle C\phi_i, \phi_i\rangle$. It then follows that $\mathrm{tr}(C_1 + C_2) = \mathrm{tr}(C_1) + \mathrm{tr}(C_2)$ whenever all three traces exist. It is also not difficult to prove that $C_1 + C_2$, AC_1, C_1A are trace class operators whenever C_1 and C_2 are and A is a bounded linear operator. Also $\mathrm{tr}(AC_1) = \mathrm{tr}(C_1A)$. Now let ϕ be a unit vector and let P_ϕ denote the operator $\psi \to \langle\psi, \phi\rangle\phi$. We see that P_ϕ is the projection onto the one-dimensional space generated by ϕ. Hence P_ϕ is a trace class operator of trace 1. If A is any bounded linear operator then $\langle P_\phi A\phi, \phi \rangle = \langle A\phi, \phi \rangle$ and for $\phi_1 \perp \phi$, $\langle P_\phi A\phi_1, \phi_1 \rangle = 0$. Hence taking ϕ to be part of a complete orthonormal set we have $\operatorname{tr}(P_\phi A) = \langle A\phi, \phi \rangle$ and hence for any projection P, $m_\phi(P) = \operatorname{tr}(P_\phi P) = \operatorname{tr}(PP_\phi)$. More generally, let B be a positive self-adjoint trace class operator with trace 1. For conciseness we shall call such operators *state operators*. Let $\phi(1)$, $\phi(2)$, ..., be the eigenvectors and $\lambda_1, \lambda_2, \ldots$, the corresponding eigenvalues of B, where $\lambda_i \geq 0$, $\sum \lambda_i = 1$. Then $\operatorname{tr}(BP) = \operatorname{tr}(PB) = \sum \langle PB\phi(i), \phi(i) \rangle = \sum \lambda_i \langle P\phi(i), \phi(i) \rangle = \sum \lambda_i m_{\phi(i)}(P)$. Thus $P \to \operatorname{tr}(BP)$ defines a state. The converse is also true. Let m be the state $m = \sum \lambda_i m_{\phi(i)}$, where $\lambda_i \geq 0$, $\sum \lambda_i = 1$, and $\phi(i) \perp \phi(j)$, $i \neq j$. Then $B = \sum \lambda_i P_{\phi(i)}$ is a state operator with eigenvectors $\phi(i)$ and corresponding eigenvalues λ_i and $m(P) = \operatorname{tr}(BP)$ for every orthogonal projection P. We have thus proved the following theorem.

Theorem 6.9. *The states may be put in one–one correspondence with the state operators in such a manner that if m and B correspond then $m(P) = \operatorname{tr}(BP)$ for every orthogonal projection P.*

For more details and applications the reader is referred to [9, 24, 25]. Note that the expectation of the bounded observable defined by the self-adjoint operator A in the mixed state $m = \sum \lambda_i m_{\phi(i)}$ is $m(A) = \sum \lambda_i \langle A\phi(i), \phi(i) \rangle = \sum \lambda_i \operatorname{tr}(AP_{\phi(i)}) = \operatorname{tr}(A \sum \lambda_i P_{\phi(i)}) = \operatorname{tr}(BA)$ where B is the state operator corresponding to m.

Let us now consider the dynamical group $t \to v_t$ in the light of our Hilbert space axiom (L-10). Let U be a unitary operator. If B is a state operator then it is easily seen that UBU^{-1} is also. Further $B \to UBU^{-1}$ maps the states of M onto M in such a way as to preserve mixtures. The same is true if U is antiunitary (antilinear, inner product preserving). On the other hand Kadison [26a] has shown that every mixture preserving one–one map of M onto M is of the form $B \to UBU^{-1}$ where U is either unitary or antiunitary, and furthermore if U_1 and U_2 define the same transformation then $U_1 = cU_2$ where $|c| = 1$. For each t choose U_t so that $v_t(B) = U_t BU_t^{-1}$. Then

$$v_{t_1+t_2}(B) = U_{t_1+t_2} BU_{t_1+t_2}^{-1} = v_{t_1}v_{t_2}(B) = U_{t_1}U_{t_2} BU_{t_2}^{-1}U_{t_1}^{-1}.$$

Thus $U_{t_1+t_2} = c(t_1, t_2)U_{t_1} U_{t_2}$ where $|c(t_1, t_2)| = 1$. In particular $U_{2t} = c(t, t)U_t^2$ so $U_t = c(t/2, t/2)U_{t/2}^2$ Since the square of an antiunitary operator is unitary it follows that U_t is unitary for all t. We have thus shown that V_t has the form $V_t(B) = U_t BU_t^{-1}$ where U_t is a unitary operator.

Lemma 6.10. (i) *If* $V_t(B) = U_t BU^{-1}$ *and* P_ϕ *is the state operator corresponding to a unit vector* ϕ, *then* $V_t P_\phi = P_{U_t(\phi)}$. (ii) *The map* $t \to |\langle U_t \phi, \psi \rangle|$ *is continuous for all vectors* ϕ, ψ.

Proof. (i) $V_t P_\phi \psi = U_t P_\phi U_t^{-1} \psi = U_t \langle U_t^{-1} \psi, \phi \rangle \phi = \langle \psi, U_t \phi \rangle U_t \phi = P_{U_t(\phi)} \psi$. (ii) By definition $t \to V_t(m)(a)$ is continuous. Therefore $t \to \text{tr}(V_t(P_\phi)P) = \text{tr}(P_{U_t(\phi)}P) = \langle U_t \phi, PU_t \phi \rangle$ is continuous for every projection P and unit vector ϕ. Let $P = P_\psi$ to get the desired result.

Part (i) of the above lemma says that the pure state P_ϕ goes into the pure state $P_{U_t(\phi)}$ after a time t. One can show that the absolute value can be dropped from part (ii) so that $t \to \langle U_t \phi, \psi \rangle$ is continuous for every ϕ, ψ. We thus have the following theorem.

Theorem 6.11. *There is a continuous one-parameter unitary group* $t \to U_t$ *such that* $V_t(B) = U_t BU_t^{-1}$ *for all t and every state operator B. Also,* V_t *corresponds to two continuous one-parameter unitary groups* U_t^1, U_t^2 *if and only if* $U_t^1 = e^{-i\alpha t} U_t^2$ *where* α *is a real constant.*

We call the one-parameter group $t \to U_t$ the *dynamical group* of our system. Applying Stone's theorem to U_t we may write it in the form $U_t = e^{-itH}$ where H is a self-adjoint operator and $-iH$ is the infinitesimal generator of U_t. The operator H which completely determines the dynamics of the system, is called the *dynamical operator*. It plays the same role in quantum mechanics that the Hamiltonian plays in classical mechanics. Like the latter it is arbitrary up to an additive constant. Since H is self-adjoint it corresponds to an observable which we call the *energy observable*. Let ϕ_0 be a unit vector in the domain of H and suppose the system is in the pure state represented by ϕ_0 at $t = 0$. Then at time t the system is in a pure state represented by $U_t(\phi_0) = e^{-itH}\phi_0$. Letting $\phi_t = e^{-itH}\phi_0$ we have $(d/dt)\phi_t = -iH\phi_t$. This equation is called *Schrödinger's equation*. It plays the role in quantum mechanics played by Hamilton's equations in classical mechanics. It is a first order differential equation whose solutions are the orbits of the pure states. A pure state P_ϕ is a *stationary state* if it is fixed in time; i.e., $V_t P_\phi = P_\phi$ for all t. It is easily seen that P_ϕ is stationary if and only if ϕ is an eigenvector of H, and thus the stationary pure states are the pure states in which the energy has a definite value with probability 1. Of course H need not have any eigenvectors at all. In the contrary case in which H has a pure point spectrum we can solve Schrödinger's equation. Let $\phi_1, \phi_2, \ldots,$ be a complete orthonormal set of eigenvectors for H

and let $\lambda_1, \lambda_2, \ldots,$ be the corresponding eigenvalues. If ψ represents a pure state we have $\psi = \sum \langle \psi, \phi_i \rangle \phi_i$ and $U_t(\psi) = e^{-itH} \psi = \sum \langle \psi, \phi_i \rangle \times e^{-it\lambda_i} \phi_i$.

An observable x is an *integral* if its probability distribution in every state is independent of time; i.e., $(V_t(m))[x(E)] = m[x(E)]$ for all $m \in M$, $E \in \mathscr{B}(R)$. It is easily seen that if x corresponds to a self-adjoint operator A, then x is an integral if and only if H and A commute. Notice the similarity between this and classical mechanics (cf. Section II) where ϕ is an integral if and only if the Poisson bracket of H and ϕ is zero. Of course, it follows that the energy observable is an integral.

With the addition of axiom (L-10) our general Axiom 3 can be given in a more convenient form. Now unitary derivables are associated with unitary operators and these in turn generate logic automorphisms on L in the natural way. Conversely, it can be shown that the type of logic automorphisms encountered in physical situations are generated by unitary operators. (The only other kind, due to antiunitary operators can be ruled out.) We now replace Axiom 3 by the following stronger axiom.

(3′) If G is the group of rigid motions on S there is a group homomorphism $W: G \to \text{auto}(L)$ (i.e., $W_{g_1 g_2} = W_{g_1} W_{g_2}$) such that
(i) $g \to m[W_g(a)]$ is continuous for every $m \in M$, $a \in L$.
(ii) $X(\Lambda g) = W_g X(\Lambda)$ for every $g \in G$, $\Lambda \in \mathscr{B}(S)$.

For $g \in G$ define $\hat{W}_g: M \to M$ by $\hat{W}_g(m)(a) = m(W_g a)$. The dynamical group V is *spatially homogeneous* if $V_t \hat{W}_g = \hat{W}_g V_t$ for all $t \in (-\infty, \infty)$, $g \in G$. This corresponds physically to the fact that one point in space is the same as another; i.e., we are in free space.

C. CONCRETE DETERMINATION OF THE QUANTUM SYSTEM

In this subsection we shall assume we have a quantum mechanical system satisfying Axioms 1, 2, and 3′. Most of the material presented here is due to Mackey [33]. It can be shown that the automorphism W_g in Axiom 3′ can be implemented by a unitary operator U_g. That is, $W_g(a) = U_g^{-1}(a)U_g$. It then follows that there exists a continuous map $\sigma: G \times G \to \{\lambda: |\lambda| = 1\}$ such that $U_{g_1 g_2} = \sigma(g_1, g_2)U_{g_1}U_{g_2}$. Thus $g \to U_g$ is a continuous unitary or projective representation of G, such that

$$X(\Lambda g) = U_g^{-1} X(\Lambda) U_g. \qquad (6.2)$$

For more details concerning the group theoretical definitions and theorems used in this subsection, the reader is referred to [30–32].

Let G be a group of rigid motions, G_1 a closed subgroup of G and G/G_1 the quotient space of right cosets. Now let A be a unitary representation of G_1 in a Hilbert space $H(A)$. Let f be a Borel function from G into $H(A)$ satisfying

$$f(g_1 g) = A_{g_1}[f(g)], \tag{6.3}$$

for every $g \in G$, $g_1 \in G_1$. Then $g \to \langle f(g), f(g) \rangle$ is constant on each coset since $\langle f(g_1 g), f(g_1 g) \rangle = \langle A_{g_1} f(g), A_{g_1} f(g) \rangle = \langle f(g), f(g) \rangle$. Thus $g \to \langle f(g), f(g) \rangle$ may be thought of as a function on G/G_1. Let v be a quasi-invariant measure on the Borel sets of G/G_1 and let H^A be the set of all Borel functions $f: G \to H(A)$ satisfying Eq. (6.3) and

$$\int_{G/G_1} \langle f(\xi), f(\xi) \rangle \, dv(\xi) < \infty. \tag{6.4}$$

Then H^A is a Hilbert space under the inner product

$$\int_{G/G_1} \langle f(\xi), g(\xi) \rangle \, dv(\xi).$$

For $g \in G$ let v_g be the measure on $\mathscr{B}(G/G_1)$ defined by $v_g(E) = v(Eg)$. Then v_g is absolutely continuous with respect to v. Let dv_g/dv be the Radon–Nikodym derivative extended in the natural way to G. Now define U_g^A on H^A by $(U_{g_1}^A f)(g) = [(dv_{g_1}/dv)(g)]^{1/2} f(gg_1)$. It is easily checked that $g \to U_g^A$ is a unitary representation of G on H^A. (U^A is a projective representation if A is projective.) U^A is called the representation of G *induced by the representation A of G_1*. It can be shown that the induced representation is independent of the quasi-invariant measure to within a unitary equivalence.

Now if $E \in \mathscr{B}(G/G_1)$, then $f \to \chi_E f$ is a projection on H^A which we denote by $P^A(E)$. It is easy to check that $E \to P^A(E)$ is a projection-valued measure on $\mathscr{B}(G/G_1)$ satisfying

$$(U_g^A)^{-1} P^A(E) U_g^A = P^A(Eg). \tag{6.5}$$

P is called a *system of imprimitivity for U^A based on G/G_1* and Eq. (6.5) is called an *imprimitivity relation*. We thus see that the representation U^A of G induced by the representation A of G_1 satisfies an imprimitivity relation. The converse of this statement is called the imprimitivity theorem [29, 30].

Theorem 6.12. *Let G_1 be a closed subgroup of G, U a unitary representation of G, and P a system of imprimitivity for U based on G/G_1. Then* *there is a representation A of G_1 (unique to within an equivalence) such that U is equivalent to the representation U^A of G induced by A. More precisely, there is a unitary operator V from $H(U)$ onto $H(U^A)$ such that $U_g = V^{-1}U_g^A V$ for every $g \in G$ and $P(E) = V^{-1}P^A(E)V$ for every $E \in \mathscr{B}(R)$. Moreover, the ring of all bounded linear operators which commute with all U_g and $P(E)$ is isomorphic to the ring of all bounded linear operators which commute with all the A_g's, $g \in G$.*

This theorem also holds for projective representations.

Now let s be a fixed element of physical spaces S and let G_s be the closed subgroup of G consisting of all $g \in G$ leaving s invariant; i.e., $(s)g = s$. Now the map $G_s g \to sg$ is a one–one map from G/G_s onto S. It can be shown that this map preserves Borel sets in both directions and in this way we may identify S with the quotient space G/G_s. Applying Eq. (6.2) we get the important result that the position σ-homomorphism is a system of imprimitivity for U_g based on G/G_s. By the imprimitivity theorem there is a representation A of G_s (unique to an equivalence) such that U and X are equivalent to the representation U^A of G induced by A and X^A, respectively. Further the ring J of bounded linear operators which commute with all U_g and $X(E)$ is isomorphic to the ring of all bounded linear operators which commute with all A_g. It is also important to note that we may now take the underlying Hilbert space H in our postulates to be $H(U^A)$. Now if J contains only the constants times the identity we say we have an *elementary* or *complete* quantum mechanical system. Most quantum mechanical systems are complete and we shall assume this property in the sequel. It now follows that the ring of bounded linear operators which commute with the A_g's consist of constants times the identity. It easily follows from this that A is an irreducible representation; i.e., there are no nontrivial closed subspaces which are left invariant under every A_g, $g \in G$. We have now proved the following important theorem.

Theorem 6.13. *In an elementary system H, U, and X are completely determined (up to an equivalence) by specifying an irreducible representation A of G_s.*

Thus if we know that a particular irreducible representation A of G_s is the relevant one then we have that $H = H(U^A)$, $U = U^A$ and $X = X^A$. Also it is easy to show that this whole structure is independent (to an equivalence) of the point $s \in S$. In fact, the choice of s corresponds to making an arbitrary choice for the origin in S. If the irreducible representation of G_s that determines the system has finite dimension j we say that the system has *spin* $(j-1)/2$.

In order to get a concrete representation for our system, let us consider a quantum mechanical system consisting of a single particle moving in R^3. Physical space S then becomes R^3. If G' is the set of transformations on R^3 which leave distances fixed it can be shown that every element of G' is a unique product of a translation, rotation, and reflection. Let G be the group of proper rigid motions; i.e., products of an element of the group of translations T with an element of the group of rotations K. Then G is a group of rigid motions on S. The closed subgroup G_s of G which leaves a point $s \in S$ invariant is just the group of rotations about s. We may suppose s is the origin and denote G_s by K. It thus becomes important to consider the group of rotations K in R^3. It is known that there is precisely one irreducible projective representation D_j of K of each finite dimension $j = 1, 2, 3, \ldots$. Thus our quantum mechanical system is determined up to an integer j, the dimension of an irreducible representation of K. Equivalently, the spin $(j-1)/2$ completely determines the quantum mechanical system.

Now suppose we know the spin (this may be determined by experiments on the particle) and thus the irreducible representation A of K. We then know that the Hilbert space is $H(U^A)$. That is, the functions $f: G \to H(A)$ such that $f(g_1 g) = A_{g_1} f(g)$ for all $g \in G$, $g_1 \in K$ and $\int_{G/K} \langle f(\xi), f(\xi) \rangle \, dv(\xi) < \infty$ where v is a quasi-invariant measure on G/K. Now it is easily seen, and in fact intuitively obvious, that we may identify G/K with T and we may obviously identify T with $S = R^3$. Thus we may consider v to be 3-dimensional Lebesgue measure and the Hilbert space becomes $L_2(R^3, H(A))$ the set of measurable functions $f: R^3 \to H(A)$ which satisfy $\int_{R^3} \langle f(\xi), f(\xi) \rangle \, dv(\xi) < \infty$. If $g = kt$, $k \in K$, $t \in T$, we see that U^A takes the form $(U_g^A f)(s) = U_{kt}^A f(s) = A_k f(k^{-1} skt)$ $= A_k f(skt) = A_k f(sg)$. We therefore have the following theorem.

Theorem 6.14. *The Hilbert space of an elementary system in R^3 with spin j may be taken to be the space $L_2(R^3, H(A))$ where A is the $2j + 1$-dimensional irreducible representation of the rotation group K and U, X have the form $U_g f(s) = A_k f(sg)$, $g = kt$, $k \in K$, $t \in T$; $X(E) f(s) = \chi_E(s) f(s)$, $E \in \mathscr{B}(R^3)$.*

Notice that $L_2(R^3, H(A))$ is equivalent to $L_2(R^3) \otimes H(A)$. Where $L_2(R^3) \otimes H(A)$ is the set of elements of the form $\sum_{i=1}^{2j+1} f_i(s) e_i$, where

$f_i = L_2(R^3)$ and $\{e_i\}$ is an orthonormal basis in $H(A)$ and

$$\langle \textstyle\sum f_i e_i, \sum h_i e_i \rangle = \sum_{i=1}^{2j+1} \langle f_i, h_i \rangle.$$

Define $U_g^0 f(s)$ on $L_2(R^3)$ by $U_g^0 f(s) = f(skt) = f(sg)$, U_g becomes $U_g^0 \times A_k$ and $X(E)$ becomes $X(E) \times I$ on $L_2(R^3) \otimes H(A)$.

Let us now find concrete representations for the position and momentum observables. Suppose we have a rectangular coordinate system $q_1(s), q_2(s), q_3(s)$. The ith coordinate observable is then $E \to q_i(X)(E) = X(q_i^{-1}(E))$. Using the spectral theorem it then easily follows that the coordinate operators are: $(q_i(X)f)(q_1, q_2, q_3) = q_i f(q_1, q_2, q_3)$, $i = 1, 2, 3$. Strictly speaking we should write $q_i(X) = q_i \times I$.

To find the momentum operators we consider one-parameter groups of transformations g_t in R^3. Now $U_{gt} = U_{gt}^0 \times A_{kt}$ where $g_t = k_t t_t$, $k_t \in K$, $t_t = T$. The infinitesimal generator is

$$B = -i\frac{d}{dt}(U_{gt})_{t=0} = -i\frac{d}{dt}(U_{gt}^0)_{t=0} \times I - iI \times \frac{d}{dt}(A_{kt})_{t=0}$$

and the momentum operator is $\hbar B$. Linear momentum in the q_1 direction corresponds to translations in the q_1 direction so that $g_t(q_1, q_2, q_3) = (q_1 + t, q_2, q_3)$. In this case $U_{gt}^0 f(q_1, q_2, q_3) = f(q_1 + t, q_2, q_3)$ and

$$Bf(q_1, q_2, q_3) = -i((d/dt)U_{gt}^0 f)_{t=0}(q_1, q_2, q_3)$$
$$= -i[(d/dt)f(q_1 + t, q_2, q_3)]_{t=0} = -i(\partial/\partial q_1)f(q_1, q_2, q_3).$$

Thus the linear momentum operators and $p_i = -i\hbar(\partial/\partial q_i) \times I$. The angular momentum about the q_3 axis corresponds to the one-parameter group of rotations k_t about the q_3 axis. Thus

$$k_t(q_1, q_2, q_3) = (q_1 \cos t + q_2 \sin t, -q_1 \sin t + q_2 \cos t, q_3).$$

Now $(U_{k_t}^0 f)(q_1, q_2, q_3) = f(q_1 \cos t + q_2 \sin t, -q_1 \sin t + q_2 \cos t, q_3)$. The *orbital angular momentum* $\omega_3 = \hbar B_3$ where B_3 is the infinitesimal generator of $U_{k_t}^0$. Since

$$(B_3 f)(q_1, q_2, q_3) = -i[(d/dt)f(q_1 \cos t + q_2 \sin t,$$
$$- q_1 \sin t + q_2 \cos t, q_3)]_{t=0}$$
$$= -i[(\partial/\partial q_1)f(q_1, q_2, q_3)(q_2)$$
$$+ (\partial/\partial q_2)f(q_1, q_2, q_3)(-q_1)]$$

we have

$$\omega_3 = i\hbar\left[q_1 \frac{\partial}{\partial q_2} - q_2 \frac{\partial}{\partial q_1}\right].$$

The operator

$$\sigma_3 = -i\hbar \frac{d}{dt} A_{k_t}\Big|_{t=0}$$

is called the *spin angular momentum* about the q_3 axis. Notice that the spin angular momentum operators are $2j + 1$ dimensional matrices called *spin matrices*. We thus have the total angular momentum about the *i*th axis $J_i = \omega_i \times I + I \times \sigma_i$, $i = 1, 2, 3$.

Let us now find the dynamical group V_t. For simplicity we shall assume that we have a spin zero system (i.e., $j = 0$). We may then take the Hilbert space to be $L_2(S)$, $(U_g f)(s) = f(sg)$ and $X(E)f = \chi_E f$. We shall also assume spatial invariance so that $V_t U_g = U_g V_t$ for all $t \in (-\infty, \infty)$, $g \in G$. Let (x, y, z) be a rectangular coordinate system for R^3 and consider the Fourier transform

$$f \to f^\sim = (2\pi)^{-1/2} \int \exp\{i(px + qy + rz)\} \cdot f(x, y, z) \, dx \, dy \, dz$$

mapping $L_2(x, y, z)$ into $L_2(p, q, r)$. Now define $U_g^\sim$ and $V_t^\sim$ on $L_2(p, q, r)$ by $U_g^\sim f^\sim = (U_g f)^\sim$ and $V_t^\sim f^\sim = (V_t f)^\sim$. Let g_1 be the translation in the x direction by $-\alpha$ units. Then

$$
\begin{aligned}
U_{q_1}^\sim f^\sim &= (U_{g_1} f)^\sim \\
&= (2\pi)^{-1/2} \int \exp\{i(px + qy + rz)\} f(x - \alpha, y, z) \, dx \, dy \, dz \\
&= e^{i\alpha p} f^\sim.
\end{aligned}
$$

Thus $U_{g_1}^\sim = e^{i\alpha p} I$ and similarly if g_2 and g_3 are translations in the y and z directions by $-\beta$, $-\gamma$ units, then $U_{g_2} = e^{i\beta q} I$, $Y_{g_3} = e^{i\gamma r} I$. Now $V_t^\sim U_g^\sim f^\sim = V_t^\sim (U_g f)^\sim = (V_t U_g f)^\sim = (U_g V_t f)^\sim = U_g^\sim V_t^\sim f^\sim$ and hence $V_t^\sim U_g^\sim = U_g^\sim V_t^\sim$. Thus $V_t^\sim$ commutes with multiplication by any function of the form $\exp\{i(\alpha p + \beta q + \gamma r)\}$. Now $\alpha \to e^{i\alpha p} I$ is a continuous one-parameter group of linear transformations on $L_2(p, q, r)$ with infinitesimal generator ip. If an operator commutes with a one-parameter group it commutes with its infinitesimal generator and hence $V_t^\sim$ commutes with multiplication by p, q, r, Since multiplication by p, q, r from a complete set of operators it follows that $V_t^\sim$ must itself be a multiplication by a function of p, q, r. Now it is easily shown that $V_t^\sim$ is a continuous one-parameter unitary group and hence $V_t^\sim = e^{-itH^\sim}$ where $H^\sim$ is a self-adjoint operator. It follows that $H^\sim$ must be multiplication by some function $h(p, q, r)$. Hence $V_t^\sim$ is multiplication by $\exp\{-ith(p, q, r)\}$.

So far we have only used the fact that V_t commutes with U_g when g

is a translation. Now let g be a rotation. Then there is an orthogonal matrix (a_{ij}) such that $(x, y, z)g = (a_{11}x + a_{12}y + a_{13}z, \ldots)$. So

$$U_{g^{-1}}\tilde{f}^{\sim} = (U_{g^{-1}}f)^{\sim} = (2\pi)^{-1/2} \int \exp\{i(px + qy + rz)\}$$
$$\times f((x, y, z)g^{-1})\, dx\, dy\, dz$$
$$= (2\pi)^{-1/2} \int \exp i\{p(a_{11}x + a_{12}y + a_{13}z) + \cdots\}$$
$$\times f(x, y, z)\, dx\, dy\, dz$$
$$= (2\pi)^{-1/2} \int \exp i\{(a_{11}p + a_{21}q + a_{31}r)x + \cdots\}$$
$$\times f(x, y, z)\, dx\, dy\, dz = \tilde{f}^{\sim}((p, q, r)g^{-1}).$$

Thus $U_g\tilde{f}^{\sim}(p, q, r) = \tilde{f}^{\sim}((p, q, r)g)$ if $g \in K$. Since $V_t^{\sim}$ and $U_g^{\sim}$ commute we have

$$(V_t^{\sim}U_g^{\sim}\tilde{f}^{\sim})(p, q, r) = \exp\{ith(p, q, r)\}(U_g^{\sim}\tilde{f}^{\sim})(p, q, r)$$
$$= \exp\{ith(p, q, r)\}\tilde{f}^{\sim}((p, q, r)g)$$
$$= U_g^{\sim}V_t^{\sim}\tilde{f}(p, q, r) = V_t^{\sim}\tilde{f}^{\sim}((p, q, r)g)$$
$$= \exp\{ith((p, q, r)g)\}\tilde{f}^{\sim}((p, q, r)g).$$

It follows that $h(p, q, r) = h((p, q, r)g)$. Now since h is independent of rotations in (p, q, r) space there is a function ρ of one real variable such that $h(p, q, r) = \rho(p^2 + q^2 + r^2)$. To find H we take the inverse Fourier transform $f \to f^{\wedge}$ to get

$$Hf(x, y, z) = [H\tilde{f}^{\sim}(p, q, r)]^{\wedge}$$
$$= (2\pi)^{-1/2} \int \exp\{-i(px + qy + rz)\}$$
$$\cdot H\tilde{f}^{\sim}(p, q, r)\, dp\, dq\, dr$$
$$= (2\pi)^{-1/2} \int \exp\{-i(px + qy + rz)\}$$
$$\cdot \rho(p^2 + q^2 + r^2)\tilde{f}^{\sim}(p, q, r)\, dp\, dq\, dr.$$

Now

$$(2\pi)^{-1/2} \int \exp\{-i(px + qy + rz)\}(p^2 + q^2 + r^2)\tilde{f}^{\sim}(p, q, r)\, dp\, dq\, dr$$

$$= -\left(\frac{\partial^2}{\partial x^2} + \frac{\partial^2}{\partial y^2} + \frac{\partial^2}{\partial z^2}\right)(2\pi)^{-1/2}$$

$$\cdot \int \exp\{-i(px + qy + rz)\}\tilde{f}^{\sim}(p, q, r)\, dp\, dq\, dr.$$

It follows from the operator functional calculus that $(Hf)(x, y, z) = \rho(-\nabla^2)f(x, y, z)$ and hence $H = \rho(-\nabla^2)$.

By applying a Galilean invariance argument which we shall not consider here it can be shown that ρ must be a linear function. Hence $H = -a\,\nabla^2 + b$ where a and b are constants and since H is determined only up to an additive constant we may assume $H = -a\,\nabla^2$. Thus a spin-zero elementary system which is Galilean invariant and spatially homogeneous is determined up to a multiplicative constant. We shall let $\mu = -\hbar^2/2a$ and call μ the *mass* of the system. So $H = -(\hbar^2/2\mu)\,\nabla^2$ describes a free spin-zero Galilean particle of mass μ_1. Let us mention that Galilean invariance merely indicates that physical laws are independent of the uniformly moving coordinate system with respect to which they are formulated. If we had insisted that our physical laws be independent of uniformly moving space *and* time coordinate axes then we would have a Lorentz invariant system. In this case we would have gotten the dynamical operator for a free spin-zero relativistic particle of rest mass μ. Thus we have derived the free Hamiltonian postulated by Schrödinger.

In the case when we do not have spatial invariance it is usually assumed that the Hamiltonian operator has the form $H = H_0 + H_i$ where H_0 is the "free" or spatially invariant Hamiltonian and H_i is the *interaction term*. The interaction term must be put in separately. If H_i is a "classical interaction" H_i is taken to be the same as in the classical system. If H_i is not classical it must be derived using physical principles.

ACKNOWLEDGMENT

I would like to thank Harry Mullikin and Steve Boyce for proofreading this paper in manuscript and correcting numerous mistakes.

REFERENCES

1. Bell, J., On the problem of hidden variables in quantum mechanics, *Rev. Modern Phys.* **38** (1966), 447–452.
2. Birkhoff, G., "Lattice Theory," Amer. Math. Soc., Providence, Rhode Island, 1948.
3. Birkhoff, G., and von Neumann, J., The logic of quantum mechanics, *Ann. Math.* **37** (1936) 823–843.

4. Bodiou, G., "Théorie dialectique des probabilitiés," Gauthier-Villar, Paris, 1964.
5. Bohm, D., and Bub, J., A proposed solution of the measurement problem in quantum mechanics by hidden variables, *Rev. Modern Phys.* **38** (1966), 453–469.
6. Bohm, D., and Bub, J., A refutation of the proof by Jauch and Piron that hidden variables can be excluded in quantum mechanics, *Rev. Modern Phys.* **38** (1966), 470–475.
7. Dunford, N., and Schwartz, J., "Linear Operators," Part I, Wiley (Interscience), New York, 1958.
8. Emch, G., and Piron, C., Symmetry in quantum theory, *J. Mathematical Phys.* **4** (1963), 469–473.
9. Fano, U., Description of states in quantum mechanics by density matrix and operator techniques, *Rev. Modern Phys.* **29** (1957), 74–93.
10. Finklestein, D., Jauch, J., Schiminovich, S., and Speiser, D., Foundations of quaternion quantum mechanics, *J. Mathematical Phys.* **3** (1963), 207–220.
11. Friedrichs, K., "Perturbation of Spectra in Hilbert Space," Amer. Math. Soc., Providence, Rhode Island, 1965.
12. Gleason, A., Measures on closes subspaces of a Hilbert space, *J. Rat. Mech. Anal.* **6** (1957), 885–893.
13. Gudder, S., Spectral methods for a generalized probability theory, *Trans. Amer. Math. Soc.* **119** (1965), 428–442.
14. Gudder, S., Uniqueness and existence properties of bounded observables, *Pacific J. Math.* **19** (1966), 81–93. [Correction, *Pacific J. Math.* **19** (1966), 588–589.]
15. Gudder, S., Hilbert space, independence and generalized probability, *J. Math. Anal. Appl.* **20** (1967), 48–61.
16. Gudder, S., Coordinate and momentum observables in axiomatic quantum mechanics, *J. Mathematical Phys.* **8** (1967), 1848–1858.
17. Gudder, S., Dispersion-free states and the existence of hidden variables, *Proc. Amer. Math. Soc.* **19** (1968), 319–324.
17a. Gudder, S., Joint distributions of observables, *J. Math. Mech.* **18** (1968), 325–335.
17b. Gudder, S., Quantum probability spaces, *Proc. Amer. Math. Soc.* **21** (1969), 296–302.
18. Guenin, M., Axiomatic foundations of quantum theories, *J. Mathematical Phys.* **7** (1966), 271–282.
19. Jauch, J., Systems of observables in quantum mechanics, *Helv. Phys. Acta* **33** (1960), 711–726.
20. Jauch, J., The problem of measurement in quantum mechanics, *Helv. Phys. Acta* **34** (1961), 293–316.
20a. Jauch, J., "Foundations of Quantum Mechanics," Addison-Wesley, Reading, Mass., 1968.
21. Jauch, J., and Misra, B., Supersymmetry and essential observables, *Helv. Phys. Acta* **34** (1961), 699–709.
22. Jauch, J., and Piron, C., Can hidden variables be excluded from quantum mechanics? *Helv. Phys. Acta* **36** (1963), 827–837.
23. Jordan, P., von Neumann, J., and Wigner, E., On an algebraic generalization of the quantum mechanical formalism, *Ann. Math.* **33** (1934), 29–64.

24. Jordan, T., Pinsky, M., and Sudarshan, E., Dynamical mappings of density operators in quantum mechanics II, *J. Mathematical Phys.* **3** (1962), 848–852.

25. Jordan, T., and Sudarshan, E., Dynamical mappings of density operators in quantum mechanics, *J. Mathematical Phys.* **2** (1961), 772–775.

26. Jost, R., "The General Theory of Quantized Fields," Amer. Math. Soc., Providence, Rhode Island, 1965.

26a. Kadison, R. V., Isometries of operator algebras, *Ann. Math.* **54** (1951), 325–338.

27. Lowdenslager, D., On postulates for general quantum mechanics, *Proc. Amer. Math. Soc.* **8** (1957), 88–91.

28. Loomis, L., On the representation of σ-complete Boolean algebras, *Bull. Amer. Math. Soc.* **53** (1947), 757–760.

29. Mackey, G., A theorem of Stone and von Neumann, *Duke Math. J.* **16** (1949) 313–325.

30. Mackey, G., Imprimitivity for representations of locally compact groups I, *Proc. Nat. Acad. Sci. U.S.A.* **35** (1949), 537–545.

31. Mackey, G., Induced representations of locally compact groups II, *Ann. Math.* **55** (1952), 101–139.

32. Mackey, G., Unitary representations of group extensions I, *Acta Math.* **99** (1958) 265–311.

33. Mackey, G., Harvard Seminar Lecture Notes. Unpublished, 1961.

34. Mackey, G., "The Mathematical Foundations of Quantum Mechanics," Benjamin, New York, 1963.

35. Mackey, G., Infinite dimensional group representations, *Bull. Amer. Math. Soc.* **69** (1963), 628–687.

36. MacLaren, M., Notes on axioms for quantum mechanics, Argonne National Laboratory, Report ANL-7065, 1965.

37. Morgenau, H., Measurement in quantum mechanics, *Ann. Physics* **23** (1963), 469–485.

38. Martin, W., and Segal, I., "Analysis in Function Space," Mass. Inst. Technol., Cambridge, Mass., 1964.

39. Misra, B., On the representation of Haag fields, *Helv. Phys. Acta* **38** (1965), 189–206.

39a. Moyal, J., Quantum mechanics as a statistical theory, *Proc. Cambridge Philos. Soc.* **45** (1949), 99–124.

39b. Mullikin, H., and Gudder, S., Measure theoretic convergences of observables and operators *J. Math. Anal. Appl.* (to appear).

40. Nahamura, M., and Umegaki, H., On von Neumann's theory of measurements in quantum statistics, *Math. Japon.* **7** (1962), 151–157.

41. Pool, J., Simultaneous observables and the logic of quantum mechanics, Ph.D. Dissertation, State University of Iowa, 1963

42. Piron, C. Axiomatique quantique, *Helv. Phys. Acta* **37** (1964), 439–468.

43. Ramsey, A., A theorem on two commuting observables, *J. Math. Mech.* **15** (1966) 227–234.

44. Rankin, B., Quantum mechanical time, *J. Mathematical Phys.* **6** (1965), 1057–1071.

45. Roberts, J., The Dirac bra and ket formalism, *J. Mathematical Phys.* **7** (1966), 1097–1104.

46. Segal, I. E., Postulates for general quantum mechanics, *Ann. Math.* **48** (1947) 930–948.
47. Segal, I. E., A non-commutative extension of abstract integration, *Ann. Math.* **57** (1953), 401–457.
48. Segal, I. E., Abstract probability spaces and a theorem of Kolmogoroff, *Amer. J. Math.* **76** (1954), 721–732.
49. Segal, I. E., "Mathematical Problems of Relativistic Physics," Amer. Math. Soc., Providence, Rhode Island, 1963.
50. Sherman, S., Non-negative observables are squares, *Proc. Amer. Math. Soc.* **2** (1951), 31–33.
51. Sherman, S., On Segal's postulates for general quantum mechanics. *Ann. Math.* **64** (1956), 593–601.
52. Sikorski, R., On the inducing of homomorphisms by mappings, *Fund. Math.* **36** (1949), 7–22.
53. Stinespring, W., Integration theorems for gages and duality for unimodular groups, *Trans. Amer. Math. Soc.* **90** (1959), 15–56.
54. Streater, R., and Wightman, A., "PCT, Spin and Statistics, and All That," Benjamin, New York, 1964.
54a. Suppes, P., The probabilistic argument for a non-classical logic for quantum mechanics, *Philos. Sci.* **33** (1966), 14–21.
55. Umegaki, H., On information in operator algebras, *Proc. Japan Acad.* **37** (1961), 459–461.
56. Umegaki, H., Conditional expectation in an operator algebra IV, *Kōdai Math. Sem. Rep.* **14** (1962), 59–85.
56a. Urbanik, K., Joint distributions of observables in quantum mechanics, *Studia Math.* **21** (1961), 117–131.
57. Varadarajan, V., Probability in physics and a theorem on simultaneous observables, *Commun. Pure Appl. Math.* **15** (1962), 189–217.
57a. Varadarajan, V., "Geometry of Quantum Mechanics," Vol. 1, Van Nostrand, Princeton, N. J., 1968.
58. von Neumann, J., On an algebraic generalization of the quantum mechanical formalism I, *Mat. Sb.* **11** (1936) 492–559. [Also in "Collected Works," Vol. 3, Pergamon Press, New York, 1961.]
59. von Neumann, J., "Mathematical Foundations of Quantum Mechanics," Princeton Univ. Press, Princeton, N. J., 1955.
60. Wicks, G., Wigner, E., and Wightman, A., Intrinsic parity of elementary particles, *Phys. Rev.* **88** (1952) 101–105.
60a. Wigner, E., On the quantum correction for thermodynamic equilibrium, *Phys. Rev.* **40** (1932) 749–759.
61. Zierler, N., Axioms for non-relativistic quantum mechanics, *Pacific J. Math.* **11** (1961), 1151–1169.
62. Zierler, N., Order properties of bounded observables, *Proc. Amer. Math. Soc.* **14** (1963), 346–351.
63. Zierler, N., On the lattice of closed subspaces of a Hilbert space, *Technical Memorandum* TM-04172, Mitre Corporation, Bedford, Mass., 1965.

RANDOM DIFFERENTIAL EQUATIONS IN CONTROL THEORY

W. M. Wonham*

CENTER FOR DYNAMICAL SYSTEMS
DIVISION OF APPLIED MATHEMATICS
BROWN UNIVERSITY
PROVIDENCE, RHODE ISLAND
AND
NASA/ELECTRONICS RESEARCH CENTER
CAMBRIDGE, MASSACHUSETTS

*Present Address: Department of Electrical Engineering, University of Toronto, Toronto 5, Canada.

I. Introduction

The primary aim of control theory is to develop mathematical models and algorithms for the design of complex dynamic systems. The goal of design is to realize a desired system function within the constraints imposed by nature, economics, and the current state of technology. In general, the function of a control system is to maintain a given set of variables within prescribed bounds. The necessity for control arises from the fact that in operation a physical system is usually subject to perturbations which cannot be exactly predicted, and thus corrected for, in advance. For this reason the presence of chance phenomena imposes a basic constraint on system performance, and it is thus appropriate to investigate control processes with the aid of stochastic models.

Study of the qualitative behavior of a system subjected to random disturbances leads naturally to various concepts of stochastic stability. An important problem is to determine effective stability criteria, as these distinguish controls which are *a priori* admissible for the application in view. In addition, a stability criterion may suggest a quantitative index on the basis of which admissible controls can be meaningfully compared. In this way one is led to the problem of optimization, that is, selection of a best possible control within an admissible class.

It seems unlikely that more than a small fraction of the control problems (stochastic or not) which arise in technology can properly be stated in rigorous variational terms. Nevertheless the study of idealized optimal processes has led to important advances in the understanding of controller structure, in particular, of its relation to the information available to the controller about past and present system behavior.

In this article we explore the foregoing questions in terms of models described by stochastic differential equations. It must be admitted that only fragmentary results are available for many of the problems raised. Furthermore, a variety of approximate methods have been found useful in engineering practice but are not yet understood rigorously from the viewpoint of differential equations. Some of these methods are considered briefly in conclusion.

In Section II some facts about stochastic differential equations are collected for ease of reference, with brief indication of how these equations arise as models for randomly perturbed dynamic systems. The simplest class of optimization problems (based on the assumption of linear dynamics and full state information) is studied in Section III; here the principal technique is dynamic programming. In Section IV we derive some known results on optimal linear filtering of signals in white Gaussian noise, and in Section V we apply these results to the problem of combined filtering and control. Section VI is concerned with control of linear systems with randomly jumping parameters. The concluding Section VII is a brief review of some miscellaneous problems of general interest, but about which little is known, together with certain approximation techniques under current development.

Notation

The following notation will be used throughout this article. Vectors x, etc. are column vectors; a prime ($'$) denotes transpose of a vector or matrix; all vectors and matrices which appear have real-valued elements. If $\phi = \phi(x)$ is a real-valued function, $\phi_x(\phi_{xx})$ denotes the vector (matrix) of first (second) partial derivatives of ϕ. R^n is Euclidean n-space; $\mathscr{B}(R^n)$ denotes the σ-algebra of Borel sets of R^n; $C_p^{(l)}(R^n)$ is the set of p-vector-valued functions on R^n with continuous derivatives up to order l. If P, Q are symmetric $n \times n$ matrices, $P > Q(P \geq Q)$ means $P - Q$ is positive (semi) definite. $|x|$ is the Euclidean norm of x and $|A|$ is the corresponding operator norm of the matrix A.

If $A(t)$ is an $n \times n$ matrix defined for $t_0 \leq t \leq t_1$, the *fundamental matrix associated with A* is the $n \times n$ matrix $\Phi(t, s)$ such that, for $t_0 \leq s, t \leq t_1$,

$$\frac{\partial \Phi(t, s)}{\partial t} = A(t)\Phi(t, s), \qquad \Phi(s, s) = I.$$

Random variables are tacitly referred to an underlying probability space $(\Omega, \mathscr{F}, \mathbf{P})$; the argument $\omega \in \Omega$ will often not be written. $\mathbf{P}$ is probability measure on a σ-algebra $\mathscr{F}$ of subsets of Ω. If $\mathscr{R}$ is a family of random variables, $\sigma(\mathscr{R})$ is the smallest σ-algebra of ω-sets with respect to which $\mathscr{R}$ is measurable. $\mathscr{E}$ denotes expectation; if $\{x(s), t \leq s \leq t_1\}$ is a stochastic process, $\mathscr{E}_{t,x}\{\cdot\}$ denotes expectation conditional on the event $x(t) = x$.

The standard Wiener process in R^m is the vector-valued process $w(t) = [w_1(t), \ldots, w_m(t)]'$, where the $w_i(t)$ are mutually independent, scalar valued, separable Wiener (Brownian motion) processes, such that

$$\mathscr{E}\{w_i(t_2) - w_i(t_1)\} = 0$$
$$\mathscr{E}\{[w_i(t_2) - w_i(t_1)]^2\} = |t_2 - t_1| \qquad (i = 1, \ldots, m),$$

for all t_1, t_2 in the interval on which $w(\cdot)$ is defined. For a discussion of the Wiener process see, e.g., Doob [10, Chapter 8].

II. Stochastic Models of Control Systems

A. Dynamic Systems Perturbed by Gaussian White Noise

The dynamic systems of control theory about which most is known are representable by ordinary vector differential equations of form

$$dx/dt = f(t, x), \qquad t_0 \leq t \leq T. \tag{2.1}$$

In (2.1) x is a vector (the system *state*) in R^n. It is assumed that f is such that (2.1) admits a unique solution $x(t; x_0)$, $t_0 \leq t \leq T$, for every initial state $x(t_0) = x_0 \in R^n$. In this chapter we assume that (2.1) represents a physical system together with its control; the control variables proper will later be distinguished explicitly.

Not all control systems which arise in applications can be represented by (2.1); we shall omit discussion of differential difference equations and partial differential equations because little information about their stochastic versions is available.

Consider next the formal stochastic differential equation

$$dx/dt = f(t, x) + G(t, x)\zeta(t), \qquad t_0 \leq t \leq T. \tag{2.2}$$

In (2.2) $G(t, x)$ is an $n \times m$ matrix-valued function of (t, x) and ζ represents m-dimensional Gaussian white noise. In writing (2.2) we usually have in mind the following situation: The drift term f of (2.1) is additively perturbed by a stationary random disturbance ζ whose power spectral density is approximately constant up to frequencies which are high relative to the time-scale of (2.1); the effective disturbance magnitude may depend on t and x; and the resulting paths $x(t)$ are continuous. The problem of making the foregoing description precise is considered later. An equation of type (2.2) was introduced by Langevin [48] in the study of Brownian motion; it was proposed as a model for a randomly disturbed dynamic system by Pontryagin *et al.* [55], and more recently, in the context of control theory, by Chuang and Kazda [7], Barrett [3], and Khazen [39].

B. Itô's Equation

To make (2.2) precise we shall consider Itô's stochastic differential equation

$$dx = f(t, x)\, dt + G(t, x)\, dw(t), \qquad t_0 \le t \le T. \qquad (2.3)$$

In (2.3) f and G are as before and w is a separable Wiener process (Brownian motion) in Euclidean m-space. Equation (2.3) was studied by Itô [27, 28] and later, under less restrictive conditions, by Doob [10], Skorokhod [63], and Dynkin [11]. Some results are summarized here for ease of reference. Equation (2.3) is interpreted as a stochastic integral equation

$$x(t) = x(t_0) + \int_{t_0}^{t} f[s, x(s)]\, ds + \int_{t_0}^{t} G[s, x(s)]\, dw(s), \qquad (2.4)$$

defined on an interval $I = [t_0, T]\ (T < \infty)$ or $I = [t_0, \infty)$. It is assumed that f and G are measurable in (t, x) for $t \in I$, $x \in R^n$; and satisfy (i) a growth condition

$$|f(t, x)| + |G(t, x)| \le k(1 + |x|), \qquad t \in I, \quad x \in R^n \qquad (2.5)$$

and (ii) a uniform Lipschitz condition

$$|f(t, x) - f(t, y)| + |G(t, x) - G(t, y)| \le k|x - y|, \qquad t \in I, \quad x, y \in R^n$$
$$(2.6)$$

for a suitable constant k. In (2.4) $x(t_0)$ is any (finite-valued) random variable independent of the increments dw. Under these conditions (2.4) determines a unique stochastic process $X = \{x(t); t \in I\}$ with the following properties:

1. The path (sample function) $x(t)$ of X is continuous on I, with probability 1.

2. If $\mathscr{E}\{|x(t_0)|^2\} < \infty$ then

$$\mathscr{E}\{\max_{t_0 \le t \le t_1} |x(t)|^2\} < \infty, \qquad t_1 \in I. \tag{2.7}$$

3. X is a separable measurable process (for the terminology see Doob [10, Chapter 1]).

4. For each $t \in I$ the random variable $x(t)$ is independent of the increments $w(t_1) - w(s), t \le s < t_1 \in I$.

5. For each $t \in I$, (2.4) holds with probability 1.

6. X is a Markov process.

It will be helpful to keep in mind the meaning of the Itô stochastic integral $\int G\, dw$. For simplicity assume (2.7) is true. Then if $t_0 \le t_1 \le \cdots \le t_v = t$, and if the t_j become dense in $[t_0, t]$ as $v \to \infty$, by definition

$$\int_{t_0}^{t} G[s, x(s)]\, dw(s) = \text{l.i.m.} \sum_{\substack{v \to \infty \\ j=0}}^{v-1} G[t_j, x(t_j)][w(t_{j+1}) - w(t_j)] \tag{2.8}$$

(For details see, e.g., Doob [10, Chapter 9], Skorokhod [63, Chapter 2], and Dynkin [11, Chapter 7].) A noteworthy feature of (2.8) is that the w-increments "point into the future" relative to the corresponding evaluations of the integrand G. As discussed in Section II.D, this fact has interesting implications for the relation between (2.3) and the Langevin equation (2.2). In addition (2.8) implies that the process

$$\gamma(t) = \int_{t_0}^{t} G[s, x(s)]\, dw(s), \qquad t \in I,$$

is a separable martingale (Doob [10, Chapter 9]); and that

$$\mathscr{E}\{\gamma(t)\} = 0,$$

$$\mathscr{E}\{\gamma(t)\gamma(t)'\} = \int_{t_0}^{t} \mathscr{E}\{G[s, x(s)]G[s, x(s)]'\}\, dt. \tag{2.9}$$

C. DIFFUSION PROCESSES

Equation (2.3) actually determines a family of Markov processes X, one for each choice of random variable $x(t_0)$; generally X will denote a fixed but arbitrary process of the family. X is a *diffusion process*, both in the intuitive sense that its main qualitative properties are shared by the classical diffusion (Brownian motion), and in the current technical sense (Dynkin [11, Chapter 5]). Let $P(t, x; s, \mathbf{B})$ be the transition function of X, defined for $t \leq s$; $t, s \in I$; $x \in R^n$ and $\mathbf{B} \in \mathscr{B}(R^n)$ (for general properties of P see, e.g., Dynkin [11, Chapter 4]). The relation between P and the coefficients f, G of (2.3) is expressed by the following basic result: if $\phi \in C_1^{(2)}(R^n)$ has compact support then

$$\lim_{s \downarrow t}(s - t)^{-1}\left[\int_{R^n} P(t, x; s, dy)\phi(y) - \phi(x)\right] = \mathscr{L}\phi(x) \qquad (2.10)$$

where

$$\mathscr{L}\varphi(x) \equiv \tfrac{1}{2} \operatorname{tr}[G(t, x)'\varphi_{xx}(x)G(t, x)] + f(t, x)'\phi_x(x) \qquad (2.11)$$

(Dynkin [11, Chapters 5, 11]).

The elliptic operator $\mathscr{L}$ is the *differential generator* of X. Under additional conditions, of smoothness and uniform ellipticity of $\mathscr{L}$, $P(t, x, s, \mathbf{B})$ itself is characterized as the unique solution of the Cauchy problem

$$u_t(t, x) + \mathscr{L}u(t, x) = 0, \qquad (t_0 \leq t \leq s, x \in R^n), \qquad (2.12)$$

$$\begin{aligned} u(s, x) &= 1, \qquad (x \in \mathbf{B}), \\ &= 0, \qquad (x \notin \mathbf{B}). \end{aligned}$$

Of importance for applications is the chain rule for computing the stochastic differential of a function along a path of X. If $\phi = \phi(t, x)$ and ϕ, ϕ_t, ϕ_x, ϕ_{xx} are continuous, the random variable $\phi(t, x(t))$ is represented by the formula

$$\phi(t, x(t)) = \phi(s, x(s)) + \int_s^t d\phi(u, x(u)) \qquad (t \geq s), \qquad (2.13)$$

Here $d\phi$ is the (Itô) stochastic differential

$$\begin{aligned} d\phi(u, x(u)) = {}&[\mathscr{L}\phi(u, x(u)) + \phi_u(u, x(u))]\, du \\ &+ \phi_x(u, x(u))'G(u, x(u))\, dw(u), \quad (2.14) \end{aligned}$$

and the stochastic integral $\int \phi_x'G\, dw$ is defined in the sense of Itô. For a proof see Gikhman and Skorokhod [22, Chapter 8]. Suppose in addition that ϕ vanishes off a compact subset of R^n. Then using the martingale property of the stochastic integral we obtain *Itô's integration ormula:*

$$\mathscr{E}_{s, x}\{\phi(t, x(t))\} = \phi(s, x) + \mathscr{E}_{s, x}\left\{\int_s^t [\mathscr{L}\phi(u, x(u)) + \phi_u(u, x(u))]\, du\right\}$$

$$(2.15)$$

More generally, recalling (2.5) and (2.7) we see that (2.15) holds if ϕ, ϕ_u, ϕ_x, ϕ_{xx} are continuous and

$$|\phi| + |\phi_u| + |x|\,|\phi_x| + |x|^2|\phi_{xx}| \leq k(1 + |x|^2) \qquad (2.16)$$

for $(u, x) \in [s, t] \times R^n$ and some constant k.

To conclude this subsection, we state a generalized version of (2.15) which will be applied in Section V. Our aim is to extend (2.15) for an $x(t)$ process which may depend on the past, in a more general (and nonMarkovian) way than does the diffusion process of (2.4). For this, let $\{w(t), t \in I\}$ be a Wiener process in R^m, as before. Let $\{\mathscr{F}_t, t \in I\}$be a family of sub-$\sigma$-algebras of $\mathscr{F}$ such that: $\mathscr{F}_{t_1} \subset \mathscr{F}_{t_2}$ if $t_1 < t_2$; $w(t)$ is $\mathscr{F}_t$-measurable $(t \in I)$; and the random variables $w(t_2) - w(t_1)$ with $t \leq t_1 < t_2$ are independent of $\mathscr{F}_t$. In place of the coefficients f, G of (2.1) introduce new functions

$$f: \quad I \times \Omega \to R^n, \qquad G: \quad I \times \Omega \to R^{n \times m},$$

with the properties:

(i) f is (t, ω)-measurable; the random variable $f(t) = f(t, \cdot)$ is $\mathscr{F}_t$-measurable $(t \in I)$ and

$$\int_{t_0}^{t_1} |f(t)|\, dt < \infty, \qquad t_1 \in I,$$

with probability 1.

(ii) G is (t, ω)-measurable; $G(t) = G(t, \cdot)$ is $\mathscr{F}_t$-measurable $(t \in I)$ and

$$\int_{t_0}^{t_1} |G(t)|^2\, dt < \infty, \qquad t_1 \in I,$$

with probability 1. Next, let $x_0(\cdot)$ be an $\mathscr{F}_{t_0}$ measurable random variable in R^n and let $x(\cdot)$ be the process in R^n with (generalized) stochastic differential

$$dx(t) = f(t)\, dt + G(t)\, dw(t), \qquad t \in I,$$

and initial value

$$x(t_0) = x_0.$$

That is,

$$x(t, \omega) = x_0(\omega) + \int_{t_0}^{t} f(s, \omega)\, ds + \int_{t_0}^{t} G(s, \omega)\, dw(s, \omega) \qquad (2.4^*)$$

To generalize (2.15) let $\phi = \phi(t, x)$ be as above; and suppose that, relative to the process (2.4*),

$$\mathscr{E}\{|\phi(t, x(t))|\} < \infty, \qquad t \in I,$$

$$\int_{t_0}^{t_1} \mathscr{E}\{|\phi_x(t, x(t))'G(t)|\}\, dt < \infty, \qquad t \in I$$

Write

$$\mathscr{L}\phi(t, x) = (1/2)\, tr[G(t)'\phi_{xx}(t, x)G(t)] + f(t)'\phi_x(t, x)$$

Subject to the foregoing assumptions the following generalization of (2.15) is valid:

$$\mathscr{E}\{\phi[t_2, x(t_2)]\} - \mathscr{E}\{\phi[t_1, x(t_1)]\}$$

$$= \mathscr{E}\{\int_{t_1}^{t_2} (\phi_t[t, x(t)] + \mathscr{L}\phi[t, x(t)])\, dt\} \qquad (2.15^*)$$

$$(t_0 \le t_1 < t_2; t_1, t_2 \in I)$$

For the definition of the stochastic integrals which occur in (2.4*) and for the proof of (2.15*) the reader is referred to Gikhman and Skorokhod [22, Chapter 9].

Further properties of diffusion processes will be introduced as they are needed. In Section II.D we turn to the question of whether Itô's equation (2.3) is suitable in applications as a precise version of (2.2).

D. ON THE RELATION BETWEEN LANGEVIN'S EQUATION AND ITÔ'S EQUATION

By Langevin's equation we mean the formal stochastic differential equation (2.2). The problem of identifying (2.2) with an Itô equation is best approached by considering the process obtained when (2.2) is simulated on an ideal analog computer. One plausible model for this situation is the following, described rather loosely. Write $A = (0, \infty)$ and, for each $\alpha \in A$, let $Z_\alpha = \{\zeta(t; \alpha), -\infty < t < \infty\}$ be a strictly stationary stochastic process in R^m with $\mathscr{E}\{\zeta(t, \alpha)\} = 0$, $\mathscr{E}\{\zeta(t, \alpha)\zeta(t + \tau, \alpha)'\}$ $= Q_\alpha(\tau)$, and Q_α continuous at $\tau = 0$. The processes Z_α are chosen to reflect the power spectral, and possibly sample, properties of whatever physical process $Z = \{\zeta(t)\}$ is postulated in (2.2). As an example suppose that ζ in (2.2) is representable as a continuous, stationary, scalar-valued process, with power spectral density approximately constant in the frequency interval $0 \le \nu \le \nu_0$, and decreasing steadily for $v > v_0$. Then Z_α might be defined as the continuous Gaussian process (Ornstein–Uhlenbeck process) wtih autocorrelation function $Q_\alpha(\tau) = (\alpha/2)e^{-\alpha|\tau|}$ where $\alpha \gtrsim 3v_0$. Alternatively, suppose the $\zeta(\cdot)$ sample paths undergo jumps at (random) times and assume (random) constant values between jumps: Z_α could be taken as a Markov step process with jump points Poisson distributed on the time axis, with parameter α, and with amplitudes chosen independently, at each jump, from a fixed probability distribution. In general Z_α is supposed to be simulable on an ideal analog computer (but infinite total average power is ruled out) and, for almost every path $\zeta(t, \alpha)$, the differential equation

$$\frac{dx}{dt} = f(t, x) + G(t, x)\zeta(t; \alpha), \qquad t \in I, \qquad (2.17)$$

is supposed to be uniquely solvable in the ordinary sense.

To match the situation where the ζ process in (2.2) has wide spectral bandwidth, impose the further requirement that, if $\Phi_\alpha(v)$ is the spectral density of Z_α, then $\Phi_\alpha(v) \to$ const. for all $v \in (-\infty, \infty)$ as $\alpha \to \infty$; this requirement holds for the examples just mentioned. Now consider the stochastic process $X_\alpha = \{x_\alpha(t); t \in I\}$ determined by (2.17). The question of interest is whether, in some sense, a limit process $X = \lim X_\alpha(\alpha \to \infty)$ exists and whether X can be represented as the solution of Itô's equation (2.3) (with possibly different coefficients f, G). If the answer is yes, then we might expect that Itô's equation (2.3) would serve as a valid precise version of (2.2). From the viewpoint of analysis (2.3) is preferable to (2.17) in certain problems of filtering, stability, and optimization where, as will be seen later, the Markov property of a diffusion process can be exploited to advantage.

Analysis of the kind just described has been performed in detail for various classes of Z_α by Wong and Zakai [70–72]. One result is the following. Let X be a scalar diffusion determined by Itô's equation

$$dx = f(t, x) \, dt + g(t, x) \, dw, \qquad t \in I = [t_0, T]; \qquad (2.18)$$

and let f, g, gg_x be continuous for $(t, x) \in I \times R^1$, and satisfy a uniform Lipschitz condition in x. Let $\{t_v^{(n)}; 1 \le v \le n\}$ $(n = 1, 2, \ldots)$ be partitions of I with $\max_v\{|t_{v+1}^{(n)} - t_v^{(n)}|\} \to 0$ $(n \to \infty)$; and define a sequence $w^{(n)}$ of polygonal approximations to w: the graph of $w^{(n)}$ on $[t_v^{(n)}, t_{v+1}^{(n)}]$ is the straight line segment joining $w(t_v^{(n)})$ to $w[t_{v+1}^{(n)}]$. (In previous notation,

$\zeta_\alpha = \zeta_n = \dot{w}^{(n)}$ is piecewise constant with amplitudes independent and Gaussian between jumps; in this model Z_n is not stationary). Let X_n be the process determined by

$$\dot{x}_n = f(t, x_n) + g(t, x_n)\dot{w}^{(n)}(t), \qquad t \in I,$$

with $x_n(t_0) = x(t_0) = \xi$, $\mathscr{E}\{\xi^4\} < \infty$. Under these conditions it is shown that

$$\text{l.i.m. } x_n(t) = y(t), \qquad (n \to \infty), \quad t \in I,$$

where the $y(t)$ process is determined by the Itô equation

$$dy = [f(t, y) + \tfrac{1}{2}g_y(t, y)g(t, y)]\, dt + g(t, y)\, dw(t),$$
$$y(t_0) = \xi. \tag{2.19}$$

Observe that (2.19) differs from (2.18) by the presence of an additional drift term contributed by the diffusion coefficient g.

The foregoing result suggests that, for a broad class of Langevin equations

$$\dot{x} = f(t, x) + G(t, x)\zeta(t), \tag{2.20}$$

an appropriate matching Itô equation is not (2.3) as written but instead

$$dx = \hat{f}(t, x)\, dt + G(t, x)\, dw$$

where $\hat{f}$ differs from f by a correction term depending on G. A suggestive determination of $\hat{f}$ makes use of an idea of Stratonovich [65]; suppose that (2.3) holds and that the analog computer represented by (2.20) evaluates the stochastic integral as

$$\int_{t_0}^t G(u, x_u)\zeta_u = \text{l.i.m.} \sum_v G\left[t_v, \frac{x(t_v) + x(t_{v+1})}{2}\right][w(t_{v+1}) - w(t_v)], \tag{2.21}$$

where $\{t_v\}$ is a partition of $[t_0, t]$ and $\max_v(t_{v+1} - t_v) \to 0$. Observe that (2.21) is a more symmetric definition of the stochastic integral than (2.8); investigation shows that a stochastic calculus based on (2.21) possesses, at least in simple cases, formal rules of manipulation (e.g., chain rule) identical with those of ordinary calculus. In other words, if $\zeta(t)$ in (2.20) is reasonably well behaved and if the computer performs ordinary (but not Itô!) calculus, we might expect to obtain the evaluation (2.21), which is formally (Stratonovich, [65])

$$\tfrac{1}{2}\int_{t_0}^t G_x(u, x(u)) \cdot G(u, x(u))\, du + \int_{t_0}^t G(u, x(u))\, dw(u). \tag{2.22}$$

In (2.22) $G = [g_{ij}]$, $G_x \cdot G$ is the vector with ith component

$$\sum_{jk} (\partial g_{ij}/\partial x_k) g_{kj},$$

and the stochastic integral is Itô's. For further discussion along these lines we refer the reader to Gikhman and Skorokhod [22, Chapter 9], Khas'minskii [38], and Clark [8].

To summarize, precise results of sufficient scope are not available, but it is plausible that the Itô counterpart of (2.20) should be taken as

$$dx = [f(t, x) + \tfrac{1}{2} G_x(t, x) \cdot G(t, x)] \, dt + G(t, x) \, dw \qquad (2.23)$$

under suitable conditions on f, G and suitable assumptions about the process $\zeta(t)$. To state the conjecture another way, Itô's equation (2.3) should be simulated on an analog computer as

$$\dot{x} = f(t, x) - \tfrac{1}{2} G_x(t, x) \cdot G(t, x) + G(t, x)\zeta(t),$$

$\zeta(t)$ being "reasonably smooth laboratory white noise." Some experimental results which bear out this conjecture are reported by Dashevskii and Lipster [9].

Little is known about the limiting diffusions obtainable if $\zeta(t)$ is impulsive in character. Results of Il'in and Khas'minskii [26] include certain conditions under which the Itô equation (2.3) represents a limiting version of the Langevin equation (2.2), as the mean impulse rate $\mu \to \infty$ and area $a \to 0$, in such a way that $\mu a \to$ constant. Some smoothness of the $\zeta(t)$ sample functions is perhaps necessary for results of the type of Wong and Zakai.

In subsequent work we employ Itô's equation without reference to its physical interpretation.

E. DYNAMIC SYSTEMS AND MARKOV STEP PROCESSES

Random discontinuous forcing terms and randomly jumping system parameters are sometimes encountered in control applications; and this suggests the introduction of stochastic models based on continuous-time Markov chains (Markov step processes). For the theory of Markov step processes we refer to Doob [10]. Let $Y = \{y(t): t \in I\}$ be a homogeneous Markov chain with state space the set of integers $S = \{1, \ldots, K\}$, and transition matrix

$$
\begin{aligned}
P(\tau) &= [P_{ij}(\tau)] \\
&= [\mathbf{P}\{y(t + \tau) = j \,|\, y(t) = i\}], \qquad t_0 \le t \le t + \tau \in I, \\
&= e^{Q\tau}. \qquad (2.24)
\end{aligned}
$$

Here $Q = [q_{ij}]$ with $q_{ij} \geq 0, j \neq i, q_{ii} = -\sum_{j \neq i} q_{ij}$; we write $q_i = -q_{ii}$.

Next suppose the dynamic system of interest is described by the stochastic differential equation

$$dx/dt = f(t, x, y(t)), \qquad t \in I, \tag{2.25}$$

where x, f are vectors in R^n. To define solutions of (2.25) assume that $f(t, x, j)$ is continuous for $(t, x) \in I \times R^n$ and satisfies growth and smoothness conditions of type (2.5), (2.6), for $j = 1, \ldots, K$. Since almost all sample functions $y(\cdot)$ are constant except for a finite number of simple jumps in any finite subinterval of I (Doob [10]), we define the paths of $x(\cdot)$ in an obvious way, joining solution arcs of (2.25) at jump points of y. The $x(t)$ sample functions so determined are then continuous with probability 1.

Let $x(t_0)$ be an arbitrary random variable independent of Y. It is easy to see that the joint process $(X, Y) = \{x(t), y(t); t \in I\}$ is Markov. Indeed if $t_0 \leq s \leq t \in I$ then $x(t)$ is determined uniquely by $x(s)$ and by $y(u)$ for $s \leq u \leq t$; and, since Y is Markov, the $y(u)$ for $u \geq s$ are independent of $y(u')$, $u' < s$, when conditioned on $y(s)$. It follows that $(x(t), y(t))$ is independent of the random variables $(x(u'), y(u'))$, $u' < s$, when conditioned on $(x(s), y(s))$.

Let $\phi(t, x, k)$ be of class $C_1^{(1)}$ in (t, x) for each $k \in S$. Using (2.24) and (2.25) it is easy to compute

$$\lim_{s \downarrow t}(s - t)^{-1}[\mathscr{E}_{t, x, k}\{\phi(s, x(s), y(s))\} - \phi(t, x, k)]$$

$$= \phi_t(t, x, k) + f(t, x, k)' \phi_x(t, x, k)$$

$$- q_k \phi(t, x, k) + \sum_{j \neq k} q_{kj} \phi(t, x, j)$$

$$\equiv \phi_t(t, x, k) + \mathscr{L}\phi(t, x, k). \tag{2.26}$$

$\mathscr{L}$ is the *generator* of (X, Y). In analogy to (2.15) it is possible to derive the formula

$$\mathscr{E}_{s, x, k}\{\varphi(t, x(t), y(t))\} = \varphi(s, x, k) + \mathscr{E}_{s, x, k}\left\{\int_s^t [\mathscr{L}\varphi(u, x(u), y(u))\right.$$

$$\left. + \varphi_u(u, x(u), y(u))] \, du\right\}. \tag{2.27}$$

Equation (2.27) will be useful in the theory of optimization.

Extensions of the foregoing model are possible, in particular to Markov chains on a continuous state space, and with inclusion of diffusive effects as in the model of Section II.B. The first extension represents little that is conceptually new, and can be carried out using the results of Doob [10, Chapter 6, Section 2]. For a diffusion component consider the stochastic differential equation

$$dx = f(t, x, y(t)) \, dt + G(t, x, y(t)) \, dw(t) \tag{2.28}$$

with $y(t)$ a Markov step process and $w(t)$ a Wiener process. Models of type (2.28) have been studied from the viewpoint of stability and control by Krasovskii and Lidskii [46]; we shall not elaborate on (2.28) here because this equation has not yet proved to be of special importance in control. In any case it is clear at this stage that (2.28) can be interpreted to define a joint process (X, Y) which is Markov, with continuous component sample functions $x(t)$. If Y is a Markov chain, as before, the generator of (X, Y) is given by

$$\begin{aligned} \mathscr{L}\phi(t, x, k) = {} & \tfrac{1}{2} \, \mathrm{tr}[G(t, x, k)'\phi_{xx}(t, x, k)G(t, x, k)] \\ & + f(t, x, k)'\phi_x(t, x, k) - q_k \phi(t, x, k) \\ & + \sum_{j \neq k} q_{kj}\phi(t, x, j). \end{aligned} \tag{2.29}$$

F. DYNAMIC SYSTEMS WITH IMPULSE DISTURBANCES

In some applications, control systems are subject to random impulsive disturbances, one example being the torques due to impingement of micrometeorites on the solar panels of a space vehicle (cf. Friedland *et al.* [19]). Stochastic differential equations which describe this situation have been studied by Itô [27] and Skorokhod [63]. Suppose as usual that the unperturbed system has differential equation

$$dx/dt = f(t, x), \qquad (t, x) \in I \times R^n, \tag{2.30}$$

and consider the stochastic equation

$$dx = f(t, x) \, dt + H(t, x) \, d\pi(t), \qquad t \in I. \tag{2.31}$$

In (2.31) H is an $n \times q$ matrix and $\{\pi(t): t \in I\}$ is a Poisson process in R^q with parameter $\lambda(t)$ and jump distribution $F(t, dz)$. More precisely, let $\lambda(\cdot)$ be a nonnegative, bounded, continuous function on I; and suppose that $F(t, \cdot)$ is a probability measure on $\mathscr{B}(R^q)$ for each $t \in I$ and that $F(\cdot, \mathbf{B})$ is continuous on bounded subintervals of I, uniformly with respect to $\mathbf{B} \in \mathscr{B}(R^q)$. Then $\pi(\cdot)$ is defined to be the process with independent, R^q-valued increments having characteristic function

$$\mathscr{E}\{\exp[iu'(\pi(t_2) - \pi(t_1))]\} = \exp\left\{\int_{t_1}^{t_2} \int_{R^q} \lambda(t)(e^{iu'z} - 1)F(t, dz)\, dt\right\}.$$

The meaning of (2.31) is that $x(\cdot)$ is determined by (2.30) except for superimposed jumps which occur according to a Poisson distribution with parameter λ; if a jump occurs at $t = \tau$, then $x(\tau + 0) - x(\tau - 0) = H(\tau, x(\tau - 0))z$, where z has distribution function F; then $x(\cdot)$ starts afresh according to (2.30) from the point $x(\tau) \equiv x(\tau + 0)$.

Skorokhod [63] considers the more general situation where λ may depend on (t, x). The precise general version of (2.31) is based on the concept of random measure; as the formalism is somewhat involved we summarize results only for the simpler case described above. Assume that f, H are continuous in (t, x) and satisfy (2.5), (2.6); in addition let

$$\int_{R^q} |z|^2 F(t, dz) < \text{constant}, \qquad t \in I,$$

and assume $x(t_0)$ is a (finite-valued) random variable independent of the $\pi(t)$ process. Under these conditions (2.31) determines a separable Markov process $X = \{x(t); t \in I\}$ with the properties:

1. The paths of X are, with probability 1, right continuous and have at most a finite number of simple jumps in every finite subinterval of I.
2, 3. X has properties 2 and 3 of Section II.B.
4. X has property 4 of Section II.B, where in property 4, π is to be read in place of w.
5. Let $\phi(t, x)$ be a scalar function, continuous together with ϕ_t, ϕ_x and such that

$$|\phi(t, x)| \leq k(1 + |x|^2), \qquad (t, x) \in I \times R^n,$$

for some constant k. The generator of the $x(t)$ process is given by

$$\mathscr{L}\phi(t, x) \equiv f(t, x)'\phi_x(t, x) - \lambda(t)\phi(t, x)$$
$$+ \lambda(t) \int_{R^q} \phi[t, x + H(t, x)z]F(t, dz) \quad (2.32)$$

Again a formula of the type of (2.15) is valid.

G. SUMMARY

The stochastic dynamic models described in this chapter provide a convenient framework for many optimization problems of Markovian type which arise naturally in control applications. The practical limitations of such models should, however, be borne in mind: Numerical results are conveniently obtainable only for systems in state space of low dimension; furthermore, the smoothness and growth conditions, imposed in the general theory on the equation coefficients, may rule out the rigorous application of results to certain control systems of practical interest. Finally, we have made the sanguine assumption that detailed information about system dynamics and system disturbances is available to the analyst. Although such is generally not the case, these idealized models will serve their purpose in illuminating the qualitative behavior of real systems.

III. Control of Linear Systems Perturbed by Gaussian or Poisson White Noise

A. INTRODUCTION

Because of their intrinsic importance and relative tractability, linear systems occupy a central place in control theory. Wiener's [70] celebrated monograph, on linear least-squares smoothing (filtering) of stationary time series, suggested to engineers the possibility of optimal design of linear control systems with respect to quadratic criteria of performance. Based on Wiener–Hopf techniques, design procedures emerged which, however, suffered from two serious defects: an inability to deal with time-variable systems, and a dependence on rather subtle complex-variable analysis which in practice restricted application to systems of low dynamic order (cf. Newton et al. [52]).

Starting about 1958, a new trend became established, stimulated partly by the rapidly increasing accessibility of digital computers and partly by the developing interest in (mainly) aerospace optimization problems of special type. Bellman's "Dynamic Programming" [4] and "Adaptive Control Processes" [5] present a computer-oriented (i.e., algorithmic) formulation of a large class of Markovian decision problems. These involve control of a (possibly nonlinear) dynamic system the behavior of which is completely characterized by a finite-dimensional vector-valued function, the *dynamic state*. Under control, the state vector evolves in time in quasi-Markovian fashion; and the optimal control law (feedback law) can in principle be computed recursively. At the same time, aerospace technology focussed attention on systems with relatively well-specified dynamic behavior, described naturally by small numbers of first-order differential equations; here the state is simply the vector of dependent variables.

These ideas have led to considerable development in the theory of optimal control of systems described by differential equations. For reasons which will appear later, dynamic programming is an especially natural framework for the study of feedback systems subject to distur- bances. A fairly complete analytical discussion can be given for linear systems, where the theory yields methods of design which have to some extent superseded the classical "frequency domain" approach men- tioned earlier.

B. LINEAR SYSTEMS WITH GAUSSIAN WHITE NOISE DISTURBANCES

Consider the formal stochastic differential equation

$$dx = A(t)x \, dt - B(t)u \, dt - C(t, u) \, dw_1 + D(t, x) \, dw_2 + E(t) \, dw_3,$$

$$t_0 \le t \le T, \quad x \in R^n$$

$$x(t_0) = x_0. \tag{3.1}$$

In (3.1) the *dynamic state* $x(t) \in R^n$; the *control vector* $u(t) \in R^m$; $w_i(t)$ $(i = 1, 2, 3)$ are independent standard Wiener processes in R^{v_i}: thus $\dot{w}_i$ formally denotes " Gaussian white noise"; x_0 is a random vector with finite covariance matrix and independent of the processes w_i.

The disturbances present are of three types: control-dependent noise $C(t, u)\dot{w}_i$, state-dependent noise $D(t, x)\dot{w}_2$, and purely additive noise $E(t)\dot{w}_3$. Control-dependent noise could arise, for instance, as a result of fluctuations in system mechanical structure or of fuel energy content. State-dependent noise could be interpreted as an internal dynamic disturbance, perhaps due to vibration modes not explicitly included in the model; additive noise could arise as an environmental disturbance which acts together with control on the basic dynamics. Although not unnatural, these interpretations are suggested only as qualitatively representative. We shall assume that

$$C(t, u) = \sum_{i=1}^{m} u_i C_i(t),$$

$$D(t, x) = \sum_{i=1}^{n} x_i D_i(t), \tag{3.2}$$

where C_i, D_i are respectively, $n \times v_1$, $n \times v_2$ matrices and u_i, x_i are the components of u and x. Then (3.1) can also be written in the suggestive form

$$\dot{x} = [A(t) + \tilde{A}(t)]x - [B(t) + \tilde{B}(t)]u + E(t)\dot{w}_3$$

where

$$\tilde{A}_{ij}(t) = \sum_{v=1}^{v_2} D_{j,\,iv}\,\dot{w}_{2v},$$

$$\tilde{B}_{ij}(t) = \sum_{v=1}^{v_1} C_{j,\,iv}\,\dot{w}_{1v};$$

that is, $\tilde{A}$ and $\tilde{B}$ are independent matrix-valued Gaussian white disturbances.

For simplicity we assume that the matrices A, B, C_i, D_i, E are piecewise continuous on $[t_0, T]$.

A variety of interesting control problems arise with (3.1). Let a quadratic cost function be given, in the form

$$L(t, x, u) = x'M(t)x + u'N(t)u. \tag{3.3}$$

We assume that M, N are piecewise continuous on $[t_0, T]$ and that $M(t) \geq 0$, $N(t) \geq \varepsilon I\,(\varepsilon > 0)$ on $[t_0, T]$. $L(t, x, u)$ is to represent the instantaneous cost when the system is in state x under control u at time t. Our system is thus a *regulator*, in which a fixed state $(x \equiv 0)$ and control $(u \equiv 0)$ represent "desired" values, and deviations due to random disturbances are to be corrected. More generally, a *tracking* problem is obtained by replacing x in (3.3) by $x - x^*(t)$, where $x^*(t)$ is a nominal, preassigned trajectory. Since this generalization merely complicates notation, it is left to the reader. The control vector $u(\cdot)$ will not be restricted *a priori* in magnitude; under these conditions the term $u'Nu$ in (3.3) is included to make the optimization problems (defined below) well-posed.

On the fixed time interval $[t_0, T]$ we consider the problem of minimizing the functional

$$J[u] = \mathscr{E}\left\{\int_{t_0}^{T} L[t, x(t), u(t)]\,dt\right\} \tag{3.4}$$

in a suitable class of controls $\mathscr{U}$. A control $u^* \in \mathscr{U}$ is *optimal* if

$$J[u^*] \leq J[u], \qquad u \in \mathscr{U}.$$

Several natural choices of $\mathscr{U}$ suggest themselves.

(i) $\mathscr{U} = \mathscr{U}_{\mathrm{OL}}$ is the class of R^m-valued piecewise continuous functions $u(t)$ on $[t_0, T]$. Such a control is *open-loop* (OL). That is, an OL control is selected once for all at the start of the control process, on the basis only of knowledge of the distribution of x_0. With OL control the actual evolution of the process (or better, the evolution of a realized path) at times $t > t_0$ has no influence on the control function input; and in general, process evolution need not and may not be observable.

(ii) At the other extreme, suppose the controller receives instantaneous information about current system state $x(t)$. It is plausible that exploiting this information would lead to superior control.

To formalize this idea, let Φ be the class of functions

$$\phi: [t_0, T] \times R^n \to R^m$$

with the properties: $\phi(t, x)$ is piecewise continuous in t for each fixed x,

$$|\phi(t, x)| \le k(1 + |x|), \qquad (t, x) \in [t_0, T] \times R^n$$

and ϕ is Lipschitz continuous in x uniformly on $[t_0, T] \times R^n$. We write $u \in \mathscr{U}_{FB}$ if

$$u(t) = \phi[t, x(t)], \qquad t_0 \le t \le T,$$

for some $\phi \in \Phi$. Observe that under these conditions (3.1), with $u = \phi$, is well-defined as an Itô equation, and that $J[u]$ is well-defined and finite (Section II.B). Such a control is called a *feedback law*. Now let $u \in \mathscr{U}_{OL}$. Setting $\phi(t, x) \equiv u(t)$ we see that $\phi \in \Phi$ and therefore $\mathscr{U}_{OL} \subset \mathscr{U}_{FB}$. Thus, on the assumption that the indicated minima exist,

$$\min\{J[u]: u \in \mathscr{U}_{FB}\} \le \min\{J[u]: u \in \mathscr{U}_{OL}\} \qquad (3.5)$$

so that optimal FB control is at least as good as optimal OL control. In typical applications the two sides of (3.5) may differ by significant amounts.

(iii) Clearly, intermediate situations are possible and are of practical importance. State information may be available only at a finite number of discrete instants or, if available continuously, may be degraded by the presence of observation noise. In some applications it is difficult or expensive, or both, to undertake measurements and the resulting trade-off between cost of control and cost of information must be examined.

In this section we determine an optimal FB control for (3.1), and compare the result with optimal OL control in a special case. Control with noisy observations is considered in Section V.

C. Optimal Feedback Control

For admissible controls we take the class $\mathscr{U}_{FB}$ defined in Section III.B. To obtain an optimal control we use the standard imbedding technique of dynamic programming together with a simple sufficient condition of the type used by Krasovskii [42] and the author [74].

Let $u = \phi(t, x)$ in (3.1) and assume $x_0 \in R^n$ is a (nonrandom) constant. From the results of Section II.B, (3.1) determines a diffusion process X_ϕ on $[t_0, T]$. Its differential generator is readily obtained after writing (3.1) in the form of (2.3). For brevity of notation we introduce mappings Γ and Δ as follows. Let $\mathscr{P}_n$ denote the class of symmetric $n \times n$ matrices. Then $\Gamma: [t_0, T] \times \mathscr{P}_n \to \mathscr{P}_m$ is defined by

$$\Gamma_{ij}(t, P) = \text{tr}[C_i(t)'PC_j(t)], \qquad t \in [t_0, T], \quad P \in \mathscr{P}_n,$$

and $\Delta: [t_0, T] \times \mathscr{P}_n \to \mathscr{P}_n$ by

$$\Delta_{ij}(t, P) = \text{tr}[D_i(t)'PD_j(t)], \qquad t \in [t_0, T], \quad P \in \mathscr{P}_n.$$

Observe that Γ and Δ are linear in P for fixed t and that $\Gamma(t, P) \geq 0$, $\Delta(t, P) \geq 0$, if $P \geq 0$. For $u \in R^m$ let $\mathscr{L}_u$ denote the elliptic operator

$$\mathscr{L}_u V(t, x) = \tfrac{1}{2}\{u'\Gamma(t, V_{xx})u + x'\,\Delta(t, V_{xx})x + \text{tr}[E(t)'V_{xx}E(t)]\}$$
$$+ [A(t)x - B(t)u]'V_x.$$

The differential generator of X_ϕ is then $\mathscr{L}_u$ with u replaced by $\phi(t, x)$.

We can now establish the following

Optimality Lemma. *Suppose there exist a control* $\phi^0 \in \Phi$, *and a function* $V: [t_0, T] \times R^n \to R^1$, *with the properties*

(i) *V, V_t, V_x, V_{xx} are continuous and, for some constant k,*

$$|V| + |V_t| + |x|\,|V_x| + |x|^2\,|V_{xx}| < k(1 + |x|^2) \qquad (3.6)$$

for all $(t, x) \in [t_0, T] \times R^n$.
(ii)

$$0 = V_t(t, x) + \mathscr{L}_{\phi^0}V(t, x) + L[t, \ x, \ \phi^0(t, x)] \qquad (3.7a)$$

$$0 \leq V_t(t, x) + \mathscr{L}_u V(t, x) + L(t, x, u) \qquad (3.7b)$$

for all $(t, x, u) \in [t_0, T] \times R^n \times R^m$;

$$V(T, x) = 0, \qquad x \in R^n. \qquad (3.7c)$$

Then ϕ^0 *is optimal.*

The proof is quite simple. Let $\phi \in \Phi$ be arbitrary and let $x(t)$, $x^0(t)$ denote respectively, the path of the process X_ϕ, X_{ϕ^0}. By Itô's integration formula (2.15) and (3.7a,c),

$$V(t, x) = -\mathscr{E}_{t,x}\left\{\int_t^T (V_t[s, x^0(s)] + \mathscr{L}_{\phi^0} V[s, x^0(s)]) \, ds\right\}$$

$$= \mathscr{E}_{t,x}\left\{\int_t^T L[s, x^0(s), \phi^0(s, x^0(s))] \, ds\right\}.$$

Similarly, using (3.7b,c), with $u = \phi(t, x)$,

$$V(t, x) = -\mathscr{E}_{t,x}\left\{\int_t^T (V_t[s, x(s)] + \mathscr{L}_\phi V[s, x(s)]) \, ds\right\}$$

$$\leq \mathscr{E}_{t,x}\left\{\int_t^T L[s, x(s), \phi(s, x(s))] \, ds\right\}.$$

In particular, setting $t = t_0$, $x = x_0$, there results

$$V(t_0, x_0) = J[u^0] \leq J[u], \qquad u \in \mathscr{U}_{FB}.$$

This inequality states that u^0 is optimal.

Observe that (3.7a,b) can be combined in the form of a functional equation of dynamic programming:

$$\min_{u \in R^m} \{V_t(t, x) + \mathscr{L}_u V(t, x) + L(t, x, u)\} = 0. \qquad (3.8)$$

The optimization problem will be solved by showing that (3.8) has a suitable solution V. In the present instance this task is straightforward. With $L(t, x, u)$ given by (3.3) the local minimization in (3.8) can be performed to give

$$\phi^0(t, x) = \tfrac{1}{2}[\tfrac{1}{2}\Gamma(t, V_{xx}(t, x)) + N(t)]^{-1}B(t)'V_x(t, x), \qquad (3.9)$$

and this substitution in (3.7a) leads to a nonlinear parabolic partial differential equation for V. Despite its unpromising appearance, this equation is explicitly solvable, by a quadratic form

$$V(t, x) = p(t) + x'P(t)x, \qquad (3.10)$$

where P is an $n \times n$ symmetric matrix. Indeed, substitution yields the equations

$$\phi^0(t, x) = K(t)x, \qquad (3.11a)$$

$$K(t) = [\Gamma[t, P(t)] + N(t)]^{-1}B(t)'P(t), \qquad (3.11b)$$

$$\dot{P} + K'\Gamma(t, P)K + \Delta(t, P) + (A - BK)'P + P(A - BK) + M + K'NK = 0,$$

$$t \in [t_0, T]$$

$$P(T) = 0 \qquad\qquad\qquad (3.11c)$$

$$p(t) = \int_t^T \text{tr}\{E(s)'P(s)E(s)\}\, ds \qquad\qquad (3.11d)$$

Equations (3.11b,c) determine a nonlinear ordinary differential equation for P. If $\Gamma \equiv 0$ (i.e., $C \equiv 0$) this reduces to the Riccati equation studied in detail by Wonham [81]. The present equation is not of Riccati type since the term $K'\Gamma(t, P)K$ is a "rational" function of P. Nevertheless (3.11b,c) can be solved by exactly the same technique (due essentially to Bellman) of quasilinearization and successive approximation, as in the paper cited. The result is that (3.8) admits exactly one solution of the form (3.10) and furthermore $P(t) \geq 0$, $t \in [t_0, T]$. It is clear that ϕ^0, V satisfy the conditions of the Optimality Lemma, and hence the function ϕ^0 given by (3.11a,b) defines an optimal feedback control $u^0 \in \mathcal{U}_{\text{FB}}$.

Two qualitative features of the optimal control are immediately evident: ϕ^0 is linear in the state x, and the matrix $K(t)$ of feedback coefficients is independent of the coefficient $E(t)$ of the purely additive noise in (2.3). On the other hand, $K(t)$ is rather sensitive to the coefficients C, D of control- and state-dependent noise. Roughly speaking, large state-dependent noise has a destabilizing effect on dynamics, and tends to increase the magnitude of the optimal "gain" K. Control-dependent noise tends to increase error by diminishing the useful effect of control, and thus large $|C|$ tends to diminish K. These effects appear with greater clarity in the investigation of stationary control, Section III.E.

To conclude this section we remark that the cost functional $J[u]$ of (3.4) can be generalized, for instance, by the addition of a quadratic term

$$\mathcal{E}\{x(T)'P_T x(T)\}, \qquad (P_T \geq 0).$$

The analysis remains essentially unchanged, with the terminal condition of (3.11c) replaced by $P(T) = P_T$.

D. OPEN LOOP CONTROL VERSUS FEEDBACK CONTROL

In this section we assume that control- and state-dependent noise are absent ($C = D \equiv 0$). The contribution to cost of purely additive noise is clear from (3.11d). We observe that the preceding discussion remains valid even if $E(t) \equiv 0$ and all stochastic features of the problem disappear: That is, the feedback control (3.11a) is optimal for the nonstochastic optimization problem obtained by replacing $\mathscr{E}\int L$ in the definition of $J[u]$ by $\int L$. In this case it is clear that open-loop and closed-loop control are equivalent: x_0 being fixed, the equation

$$\dot{x} = A(t)x - B(t)\phi^0(t, x), \qquad t \in [t_0, T],$$
$$x(t_0) = x_0,$$

uniquely determines $x(t)$. If we define the open loop control

$$u(t) \equiv \phi^0[t, x(t)], \qquad t \in [t_0, T],$$

then, obviously, $x(t)$ is recovered by solving

$$\dot{x} = A(t)x - B(t)u(t), \qquad t \in [t_0, T],$$
$$x(t_0) = x_0.$$

Thus in the nonstochastic case equality is realized in (3.5).

An essentially "stochastic" feature of the control problem is the strict inequality in (3.5) which results, in general, when noise is present. The comparison is made explicit by the following observations.

(i) Let $K(t) = N(t)^{-1}B(t)'P(t)$ be defined as in (3.11b) and let $\mathscr{A}(t, t_0)$ be the fundamental matrix associated with $A(t) - B(t)K(t)$. The optimal open-loop control is then

$$u_0(t) = K(t)\mathscr{A}(t, t_0)x_0 \qquad (3.12)$$

Observe that $u_0(t)$ does not depend on the coefficient $E(\cdot)$ of the additive noise.

(ii) Let $\mathscr{A}(s, t)$ be the fundamental matrix associated with $A(s)$. Define the matrix

$$R(t) = \int_t^T \mathscr{A}(s, t)'(PBN^{-1}B'P)(s)\mathscr{A}(s, t)\, ds. \qquad (3.13)$$

Then, if u^0 denotes optimal feedback control, it can be shown that

$$J[u_0] - J[u^0] = \int_{t_0}^T \operatorname{tr}\{E(s)'R(s)E(s)\}\, ds. \qquad (3.14)$$

Trivial special cases apart, the difference $J[u_0] - J[u^0]$ in (3.14) is a positive number and thus strict inequality occurs in (3.5) whenever the noise coefficient $E(\cdot)$ is not identically zero. This difference increases (roughly speaking) as $|E|^2$, and increases also with the duration of the control process. The latter rate of increase is exponential if, for instance, the parameter matrices A, B, E, M, N are constant and A has an eigenvalue α with Re $\alpha > 0$. The rather obvious moral is, "No open-loop control can stabilize an inherently unstable system, and feedback is crucial for acceptable long-term operation."

The verification of (3.12)–(3.14), while straightforward, is somewhat lengthy and will be omitted.

E. Optimal Stationary Control

In this subsection we discuss the problem of optimal control under stationary conditions, of a time-invariant system of the form (3.1). In physical terms, our system has constant (or very slowly varying) parameters, and the control interval $T - t_0$ is large compared to the relaxation times of system dynamics. Since in (3.4) we have, typically, that $J[u] \to \infty$ as $T - t_0 \to \infty$, it is natural to modify the control problem in such a way that the new cost functional resembles the average long-term cost, say

$$\lim_{T-t_0 \to \infty} \mathscr{E} \left\{ (T - t_0)^{-1} \int_{t_0}^{T} L[t, x(t), u(t)] \, dt \right\}.$$

Clearly an ergodic property is involved here: it turns out to be convenient to work directly in terms of ergodic measures. Afterwards, it will be possible to show that, under suitable conditions, the optimal feedback control for the finite-time problem (Section III.C) tends to the optimal control for the stationary problem, in the limit as $T - t_0 \to \infty$.

From now on we assume that in (3.1)–(3.3) the parameter matrices A, B, E, M, N and C_i, $D_j (i = 1, \ldots, m; j = 1, \ldots, n)$ are constant. To avoid possible degeneracy of the ergodic measures introduced later, we assume also that $EE' > 0$: this guarantees that the transition function of each diffusion process $X_\phi (\phi \in \Phi)$ assigns, for $t > 0$, positive probability to every nonempty open set in R^n.

It will be convenient to use the terminology of constant linear systems (cf. Wonham [78]). The ordered pair of matrices (A, B) is *controllable* if the $n \times mn$ matrix

$$Z(A, B) = [B, AB, \ldots, A^{n-1}B],$$

is of rank n (i.e., the range of Z is all of R^n). Let the minimal polynomial $\alpha(\lambda)$ of A be factored as $\alpha(\lambda) = \alpha^+(\lambda)\alpha^-(\lambda)$, where $\alpha^\pm$ are polynomials such that all zeros of $\alpha^+(\lambda)$ lie in the closed right-half complex plane and all zeros of $\alpha^-(\lambda)$ lie in the open left-half plane. Let

$$E_A^+ = \{x : x \in R^n, \alpha^+(A)x = 0\}.$$

The pair (A, B) is *stabilizable* if Range $\{Z(A, B)\} \supset E_A^+$. The pair (A, B) is stabilizable if and only if there exists an $m \times n$ matrix K such that $A + BK$ is stable. Let F be an $f \times n$ matrix. The pair (F, A) is *observable* if (A', F') is controllable.

In this subsection we write Φ for the class of functions $\phi : R^n \to R^m$ with the property that

$$|\phi(x) - \phi(y)| < k_1|x - y|, \qquad x, y \in R^n,$$

for some constant k_1 (depending on ϕ). As a consequence

$$|\phi(x)| < k_2(1 + |x|), \qquad x \in R^n,$$

for some constant k_2. Consider the homogeneous diffusion process X_ϕ in R^n determined by the Itô equation

$$dx = Ax \, dt - B\phi(x) \, dt - C[\phi(x)] \, dw_1 + D(x) \, dw_2 + E \, dw_3, \qquad t \geq 0,$$

$$x(0) = x_0.$$

X_ϕ is said to have an *ergodic probability measure* μ_ϕ if μ_ϕ is a probability measure on the Borel sets of R^n and if, for every Borel set $\mathbf{B} \subset R^n$,

$$\mathbf{P}\{x_0 \in \mathbf{B}\} = \mu_\phi(\mathbf{B}) \text{ implies } \mathbf{P}\{x(t) \in \mathbf{B}\} = \mu_\phi(\mathbf{B})$$

for all $t > 0$. The restriction $EE' > 0$ implies that if μ_ϕ exists it is unique (Khas'minskii [37]). Under these conditions X_ϕ is *ergodic*. If $L(x)$ is a real-valued, Borel-measurable and μ_ϕ-integrable function on R^n, then by the individual ergodic theorem,

$$\lim_{t \to \infty} t^{-1} \int_0^t L[x(s)] \, ds = \int_{R^n} L(x)\mu_\phi(dx)$$

with probability 1 (see, e.g., Rozanov [60, Chapter 4]).

Given $\phi \in \Phi$, an ergodic measure μ_ϕ need not exist: In fact, existence of μ_ϕ is a stability property which reflects a tendency of the path $x_\phi(\cdot)$ to spend most of the time in a sufficiently large ball in R^n. More precisely, μ_ϕ exists if and only if, for every nonempty bounded open set $S \subset R^n$ with smooth boundary ∂S, the expected passage time

$$\mathscr{E}_x\{\min t : x_\phi(t) \in \partial S\} < \infty \tag{3.15}$$

for all $x \in R^n$. If (3.15) holds, X_ϕ is *positive recurrent*.

Even if μ_ϕ exists, it need not have a finite moment of preassigned order. We shall say that $\phi \in \Phi$ is *admissible*, and write $\phi \in \Phi_a \subset \Phi$, if μ_ϕ exists and

$$\mathscr{E}_\phi\{|x|^2\} \equiv \int_{R^n} |x|^2 \mu_\phi(dx) < \infty. \tag{3.16}$$

It can be shown (Wonham [76]) that (3.16) is equivalent to a strengthened version of (3.15):

$$\mathscr{E}_x\left\{\int_0^\tau |x_\phi(t)|^2 \, dt\right\} < \infty, \qquad x \in R^n,$$

where τ is the first passage time to ∂S.

As in Section III.C, let $\mathscr{L}_u$ denote the operator

$$\mathscr{L}_u V(x) \equiv \tfrac{1}{2}\{u'\Gamma(V_{xx})u + x' \Delta(V_{xx})x + \operatorname{tr}(E'V_{xx}E)\} + (Ax - Bu)'V_x \tag{3.17}$$

Then $\mathscr{L}_\phi \equiv \mathscr{L}_{\phi(x)}$ is the differential generator of X_ϕ.

The following criterion for admissibility of ϕ is a corollary of Theorems 2 and 3.1 of Wonham [75, 76], respectively.

Lemma 3.1 (Stability). *Suppose there exist a number $\rho > 0$ and a real-valued function $V(x)$ defined for $|x| > \rho$, such that*

(i)   *V, V_x, V_{xx} are continuous $(|x| > \rho)$,*
(ii) *$V(x) \to \infty$ as $|x| \to \infty$, and*
(iii) *$\mathscr{L}_\phi V(x) \le -|x|^2 \ (|x| > \rho)$.*

Then μ_ϕ exists and (3.16) *is true.*

It is now possible to state a meaningful problem of optimization. Let $M > 0$, $N > 0$. *With*

$$L(x, u) = x'Mx + u'Nu$$

find $\phi^0 \in \Phi_a$ such that

$$\mathscr{E}_{\phi^0}\{L[x, \phi^0(x)]\} \le \mathscr{E}_\phi\{L[x, \phi(x)]\}, \qquad \phi \in \Phi_a.$$

The feedback control ϕ^0 is *optimal.*

The following discussion is an easy generalization of that given in Wonham [77] and for this reason proofs will only be sketched. The first step is to verify that Φ_a is nonempty. This is true provided (A, B)

is stabilizable and the intensities of state-dependent noise and control-dependent noise are sufficiently small. In the following, K denotes a constant $m \times n$ matrix.

Lemma 3.2 (Existence of an admissible control). *An admissible control ϕ exists if*

$$\inf_K \left| \int_0^\infty e^{t(A-BK)'}[K'\Gamma(I)K + \Delta(I)]e^{t(A-BK)} \, dt \right| < 1. \qquad (3.18)$$

Furthermore, ϕ can be chosen linear in x:

$$\phi(x) = Kx.$$

The result, an application of Lemma 3.1, is proved by setting $\phi(x) = Kx$, and seeking a quadratic function $V(x) = x'Qx$ ($Q > 0$) such that

$$\mathcal{L}_\phi V(x) = q - |x|^2, \qquad x \in R^n,$$

for a suitable number q.

Next we find a criterion of optimality. The following result plays the role of the optimality lemma of Section III.C.

Lemma 3.3 (Optimality). *Suppose there exist a function $\phi^0 \in \Phi_a$, a number λ and a real-valued function $V(x)$ on R^n, with the properties*

(i)   *V, V_x, V_{xx} are continuous on R^n, and*

$$|V(x)| + |x| |V_x(x)| + |x|^2 |V_{xx}(x)| < k(1 + |x|^2), \qquad x \in R^n, \quad (3.19)$$

for some constant k.

(ii)

$$\lambda = \mathcal{L}_{\phi^0} V(x) + L(x, \phi^0(x)) \qquad (3.20a)$$

$$\lambda \leq \mathcal{L}_u V(x) + L(x, u) \qquad (3.20b)$$

for all $(x, u) \in R^n \times R^m$.

Then ϕ^0 is optimal.

To prove the lemma we note first that if V is any function which satisfies (i) and if $\phi \in \Phi_a$, then $\mathcal{L}_\phi V(x)$ is μ_ϕ-integrable and

$$\int_{R^n} \mu_\phi(dx)\mathcal{L}_\phi V(x) = 0. \qquad (3.21)$$

Equation (3.21) follows easily (see Wonham [77]) from the fact that μ_ϕ is an invariant measure for X_ϕ. Next (3.20a), (3.21) yield

$$\lambda = \mathscr{E}_{\phi^0}\{L[x, \phi^0(x)]\}. \tag{3.22}$$

Let $\phi \in \Phi_a$ be arbitrary. By (3.20b) and (3.21)

$$\lambda \le \mathscr{E}_\phi\{\mathscr{L}_\phi V(x) + L[x, \phi(x)]\} = \mathscr{E}_\phi\{L[x, \phi(x)]\}. \tag{3.23}$$

Relations (3.22), (3.23) imply that ϕ^0 is optimal.

With these results it is possible to prove:

Theorem 3.4. *If (A, B) is stabilizable, if $M > 0$, $N > 0$, and if Γ, Δ satisfy (3.18), then an optimal control exists, of the form*

$$\phi^0(x) = K^0 x. \tag{3.24}$$

Furthermore

$$K^0 = [\Gamma(P^0) + N]^{-1} B' P^0, \tag{3.25}$$

where P^0 is the unique solution, in the class of positive semidefinite matrices, of the equation

$$M + A'P + PA + \Delta(P) - PB[\Gamma(P) + N]^{-1} B'P = 0. \tag{3.26}$$

The minimum cost is

$$\mathscr{E}_{\phi^0}\{L(x, \phi^0)\} = \operatorname{tr}(E'PE). \tag{3.27}$$

For the proof we determine

$$V(x) = x'Px, \tag{3.28}$$

such that $P > 0$ and V satisfies the conditions of Lemma 3.3. Observe that (3.20a,b) yield an equation of Bellman's type:

$$\min\{\mathscr{L}_u V(x) + L(x, u)\} = \lambda. \tag{3.29}$$

With the substitution (3.28) in (3.29) we at once obtain (3.24)–(3.26). To show that (3.26) has a solution we rewrite it as

$$(A - BK)'P + P(A - BK) + K'\Gamma(P)K + \Delta(P) + M + K'NK = 0, \tag{3.30}$$

$$K = [\Gamma(P) + N]^{-1} B'P. \tag{3.31}$$

Next observe that, if $A - BK$ is stable, (3.30) is equivalent to

$$P = T_K(P) + R_K, \tag{3.32}$$

where

$$T_K(P) = \int_0^\infty e^{t(A-BK)'}[K'\Gamma(P)K + \Delta(P)]e^{t(A-BK)}\,dt, \qquad (3.33)$$

$$R_K = \int_0^\infty e^{t(A-BK)'}(M + K'NK)e^{t(A-BK)}\,dt. \qquad (3.34)$$

The remaining argument is an easy generalization of Wonham [77]. For arbitrary symmetric P,

$$-|P|\,T_K(I) \le T_K(P) \le |P|\,T_K(I);$$

hence, by (3.18), there exists $K = K_1$ such that $T_1 = T_{K_1}$ is a contraction mapping, and then (3.32) has a unique solution $P = P_1$. In general let $T_v = T_{K_v}$, $R_v = R_{K_v}$,

$$P_v = T_v(P_v) + R_v,$$
$$K_{v+1} = [\Gamma(P_v) + N]^{-1}B'P_v, \qquad (v = 1, 2, \ldots).$$

It is easily shown (cf. Wonham [77]) that the sequences $\{P_v\}$, $\{K_v\}$ are well-defined and that $0 < P_{v+1} \le P_v$. This implies that $P^0 = \lim P_v$ $(v \to \infty)$ exists, and K^0 is defined by (3.25). Since $P^0 \ge 0$ and $M > 0$, it results from (3.30) that actually $P^0 > 0$ and that $A - BK^0$ is stable. Now $V(x) = x'P^0 x$ satisfies

$$\mathcal{L}_{\phi^0} V(x) = \text{tr}(E'PE) - x'(M + K^{0'}NK^0)x, \qquad x \in R^n;$$

and it follows by Lemmas 3.1 and 3.3 that $\phi^0 \in \Phi_a$, and ϕ^0 is optimal. The evaluation (3.27) results from (3.22). The asserted uniqueness of P^0 is a consequence of the minimum property of a solution of (3.26) (cf. Wonham [81]). The proof of Theorem 3.4 is complete.

The assumption $M > 0$ can be weakened slightly. By a somewhat more complicated argument (cf. Wonham [81]) it can be shown that Theorem 3.4 remains true if $M \ge 0$ and the pair $(\sqrt{M}, A)$ is observable.

Example. The following artificial example is of interest because it illustrates explicitly the effect of state space dimension and the role of state- and control-dependent noise. Let $m = d_1 = d_2 = n$ and

$$A = aI, \qquad\qquad B = bI,$$
$$C(u) = c\,|u|\,I, \qquad D(x) = d\,|x|\,I,$$
$$M = \mu I, \qquad\qquad N = vI,$$

where a, and b, c, d, μ, $v > 0$, are scalars. Although C, D are not linear in u, x the spherical symmetry yields similar results, and we have

$$\Gamma(P) = c^2 \operatorname{tr}(P)I, \quad \Delta(P) = d^2 \operatorname{tr}(P)I.$$

With $K = kI$ for a scalar k, (3.18) yields

$$\inf_{k > a/b} \frac{(k^2 c^2 + d^2)n}{2(bk - a)} < 1$$

or

$$\theta \equiv (2na + n^2 d^2)c^2/b^2 < 1. \tag{3.35}$$

From (3.26) the matrix $P^0 = p^0 I$ is determined as

$$p^0 = \tfrac{1}{2}(1 - \theta)^{-1}[\theta(v/nc^2) + (\mu nc^2/b^2) \\ + \{[\theta(v/nc^2) + (\mu nc^2/b^2)]^2 + 4(1 - \theta)\mu v/b^2\}^{1/2}] \tag{3.36}$$

The optimal control is then $\phi^0(x) = k^0 x$, with

$$k^0 = (nc^2 p^0 + v)^{-1} bp^0. \tag{3.37}$$

From these expressions one may note several interesting qualitative effects. Inequality (3.35) provides a condition for stabilizability, in the sense of existence of a control such that a corresponding ergodic measure exists and has a finite second moment. The condition is satisfied for all c only if $2a \le -nd^2$, i.e., the "uncontrolled linear dynamics are sufficiently stable" in the ordinary sense. Otherwise stabilization is possible only for c sufficiently small. If $2a < -nd^2$ and $c \to \infty$ then

$$p^0 \to \mu |2a + nd^2|^{-1}, \quad k^0 \to 0,$$

and control is, in effect, useless. If control-dependent noise is absent ($c = 0$), (3.35) shows that stabilization by linear feedback is possible for all levels of state-dependent noise and by (3.36),

$$p^0 \sim vnd^2/b^2, \quad k^0 \sim nd^2/b, \quad (d \to \infty).$$

In general, however, stabilization by linear feedback may not be possible for large state-dependent noise even if control is noise-free: For an example of such instability see Wonham [77], and for further discussion of the question of stabilizability see Krasovskii [43].

We remark finally that, in the general case, (3.18) is automatically satisfied if both control- and state-dependent noise are absent ($C(u) \equiv 0$,

$D(x) \equiv 0$). This is the "standard" problem of the stationary linear regulator. The solution always exists if (A, B) is stabilizable, $N > 0$, and $(\sqrt{M}, A)$ is observable; furthermore, the matrix $A - BK^0$ is stable.

F. ASYMPTOTIC BEHAVIOR OF CONTROL FOR LARGE TIME-INTERVALS

From the discussion of Sections III.C,E one would expect that, under suitable conditions the optimal feedback control matrix $K(t)$, obtained for the functional (3.4), would tend to the stationary control matrix K^0 corresponding to the functional $\mathscr{E}_\phi\{L(x, \phi)\}$, as the control interval $T - t_0 \to \infty$. This is true provided the solution $P(t)$ of (3.11) tends to the solution P^0 of (3.26) as $t \to -\infty$. A proof of this limit property for the case $\Gamma \equiv 0$ (i.e., state-dependent noise only) is given by Wonham [81], and can be generalized without difficulty to (3.11) and (3.26). To summarize, the limit property holds provided the parameter matrices A, B, E, M, N, and C_i, D_j $(i = 1, \ldots, m; j = 1, \ldots, n)$ are constant, and if the conditions of Theorem 3.4 are satisfied.

G. POISSON IMPULSIVE DISTURBANCES

The analysis of Section III.C is easily extended to include the presence of Poisson impulsive disturbances together with the Gaussian white noise represented by the differentials dw_i in (3.1). To make the extension consider the formal stochastic differential equation

$$dx = A(t)x \, dt - B(t)u \, dt - C(t, u) \, dw_1 - \tilde{C}(t, u) \, d\pi_1$$
$$+ D(t, x) \, dw_2 + \tilde{D}(t, x) \, d\pi_2 + E(t) \, dw_3 + \tilde{E}(t) \, d\pi_3,$$

$$t_0 \leq t \leq T, \quad (3.38)$$

$$x(t_0) = x_0$$

In (3.38) all terms apart from those involving $d\pi_i$ are defined as in (3.1); the $\pi_i(\cdot)$ are Poisson processes of the type described in II.F and π_i is assumed to be independent of π_j $(j \neq i)$, the w_i $(i = 1, 2, 3)$, and x_0. In addition

$$\tilde{C}(t, u) = \sum_{i=1}^{m} u_i \tilde{C}_i(t),$$

$$\tilde{D}(t, x) = \sum_{i=1}^{n} x_i \tilde{D}_i(t).$$

In the notation of Section II.F we associate with the π_i nonnegative rate functions $\lambda_i(t)$ and jump distributions $F_i(t, dz)$. It will be assumed that, for $t \in I$ and $i = 1, 2, 3$,

$$\int zF_i(t, dz) = 0,$$
$$\int zz'F_i(t, dz) = Q_i. \tag{3.39}$$

Under these conditions, according to (2.32) the operator $\mathscr{L}_u$ of Section III.C is augmented by the addition of terms

$$-[\lambda_1(t) + \lambda_2(t) + \lambda_3(t)]V(t, x)$$
$$+ \lambda_1(t) \int V[t, x - \tilde{C}(t, u)z]F_1(t, dz)$$
$$+ \lambda_2(t) \int V[t, x + \tilde{D}(t, x)z]F_2(t, dz) \tag{3.40}$$
$$+ \lambda_3(t) \int V[t, x + \tilde{E}(t)z]F_3(t, dz).$$

In particular if $V(t, x) = p(t) + x'P(t)x$ then the expression (3.40) reduces to

$$u'\tilde{\Gamma}(t, P)u + x' \tilde{\Delta}(t, P)x + \lambda_3(t) \, \text{tr}\{\tilde{E}(t)'P(t)\tilde{E}(t)Q_3(t)\}.$$

Here $\tilde{\Gamma}: [t_0, T] \times \mathscr{P}_n \to \mathscr{P}_m$ and $\tilde{\Delta} = [t_0, T] \times \mathscr{P}_n \to \mathscr{P}_n$ are defined by

$$[\tilde{\Gamma}(t, P)]_{ij} = \lambda_1(t) \, \text{tr}\{\tilde{C}_i(t)'P(t)\tilde{C}_j(t)Q_1(t)\}, \qquad (i, j = 1, \ldots, m),$$
$$[\tilde{\Delta}(t, P)]_{ij} = \lambda_2(t) \, \text{tr}\{\tilde{D}_i(t)P(t)\tilde{D}_j(t)Q_2(t)\}, \qquad (i, j = 1, \ldots, n).$$

The remainder of the analysis proceeds exactly as in Section III.C, leading to equations identical with 3.11, except that Γ, Δ are replaced, respectively by $\Gamma + \tilde{\Gamma}$, $\Delta + \tilde{\Delta}$, and (3.11d) is replaced by

$$p(t) = \int_t^T \text{tr}\{E(s)'P(s)E(s) + \lambda_3(s)\tilde{E}(s)'P(s)\tilde{E}(s)Q_3(s) \, ds\}.$$

The discussion of Sections III.D and III.F can also be extended in an obvious way; however, the investigation of stationary control, although formally similar to that of Section III.E, has not yet been carried out in detail and will be omitted here.

IV. Optimal Linear Filtering

A. INTRODUCTION

In this section we study optimal linear filtering along lines proposed by Kalman [34, 35] and Kalman and Bucy [36]. This approach, which amounts to sequential linear regression, represents a practical simplification of the fundamental results of Wiener [70]. Briefly stated, the essence of the present method is that an explicit dynamic model of the basic stochastic processes is assumed at the start and thus one avoids the difficult problem of spectral factorization. The present method yields a dynamic specification of the optimal filter which is readily simulated by computer.

The theory will be applied in Section V to the problem of optimal control of a system with linear dynamics, when the state information available to the controller is perturbed by additive Gaussian white noise.

B. OPTIMAL FILTERING—PRELIMINARIES

We start by recalling some facts from elementary statistics. Let $x = x(\omega)$, $y = y(\omega)$ be random vectors with real-valued components and finite second moments. Write

$$
\begin{aligned}
\mathcal{E}\{x\} &= m_x, \qquad \mathcal{E}\{y\} = m_y, \\
\mathcal{E}\{(x - m_x)(x - m_x)'\} &= P, \\
\mathcal{E}\{(x - m_x)(y - m_y)'\} &= Q, \\
\mathcal{E}\{(y - m_y)(y - m_y)'\} &= R.
\end{aligned}
\tag{4.1}
$$

Consider the problem: Given the value $y(\omega)$, what is the optimal estimate of $x(\omega)$? Let $\hat{x}(\omega)$ be the corresponding estimate. In the special but important case where: (i) the estimate desired is *unbiased*, i.e., $\mathcal{E}\{\hat{x}\} = \mathcal{E}\{x\}$; (ii) the estimate is *linear*: $\hat{x}(\omega) = Ay(\omega) + b$ (A, b are constant); (iii) the criterion of optimality is *minimum weighted expected squared error*:

$$
\mathcal{E}\{(x - \hat{x})'M(x - \hat{x})\} = \text{minimum}, \qquad (M > 0);
$$

then the solution is well known:

$$
\hat{x}(\omega) = m_x + QR^{-1}[y(\omega) - m_y].
\tag{4.2}
$$

The covariance of error is

$$\mathscr{E}\{(x - \hat{x})(x - \hat{x})'\} = P - QR^{-1}Q'. \tag{4.3}$$

Here we have assumed $R > 0$; if $\det R = 0$, R^{-1} can be replaced by the pseudo-inverse.

If x, y are Gaussian random variables then the estimate (4.2) is also the conditional expectation of x given y: $\hat{x} = \mathscr{E}\{x \mid y\}$. In that case the estimate (4.2) is optimal in the class of all estimators subject to (i) and (iii). A direct computation from Gaussian densities shows that (4.3) also gives the *a posteriori* (i.e., conditional) covariance of x:

$$\mathscr{E}\{(x - \hat{x})(x - \hat{x})' \mid y\} = P - QR^{-1}Q'. \tag{4.4}$$

Observe that in the Gaussian case the *a posteriori* covariance does not depend on y.

Now let X, Y be the Gaussian processes determined by linear Itô equations

$$dx(t) = A(t)x(t)\,dt + B(t)\,dw(t), \tag{4.5a}$$

$$dy(t) = C(t)x(t)\,dt + D(t)\,dw(t), \tag{4.5b}$$

defined for $t_0 \le t \le T$; with initial conditions

$$x(t_0) = x_0, \qquad y(t_0) = 0. \tag{4.6}$$

In (4.5), the parameter matrices are assumed to be continuous on $[t_0, T]$; for some $\varepsilon > 0$, $D(t)\,D(t)' \ge \varepsilon I$, $t_0 \le t \le T$; $\{w(t, \omega)\}$ is a d-dimensional standard Wiener process; and in (4.6), x_0 is a Gaussian random variable independent of the $w(t)$ process. Equation (4.5) is assumed to describe a linear dynamic process perturbed by Gaussian white noise, and with state $x(\cdot)$ observed via a noisy linear channel with output $y(\cdot)$. The problem is to estimate $x(t)$ given $y(s)$ for $t_0 \le s \le t$; the parameter matrices A, B, C, D, and the mean and covariance of x_0 are assumed to be known.

Let

$$\mathscr{Y}_t = \sigma\{y(s): t_0 \le s \le t\}.$$

As the estimate of $x(t)$ we adopt[1]

$$\hat{x}(t) = \mathscr{E}\{x(t) \mid \mathscr{Y}_t\}. \tag{4.7}$$

[1] It will be unnecessary to distinguish between "versions" of $\hat{x}$.

Let $P(t)$ be the conditional covariance

$$P(t) = \text{cov}\{x(t) \mid \mathscr{Y}_t\}$$
$$= \mathscr{E}\{(x(t) - \hat{x}(t))(x(t) - \hat{x}(t))' \mid \mathscr{Y}_t\}, \qquad (4.8)$$

and write, for $t \geq s$,

$$P(t \mid s) = \text{cov}\{x(t) \mid \mathscr{Y}_s\}. \qquad (4.9)$$

Finally let $\Phi(t, s)$ be the fundamental matrix associated with $A(\cdot)$.
We begin by observing that

$$x(t) = \Phi(t, s)x(s) + \int_s^t \Phi(t, \tau)B(\tau)\, dw(\tau) \qquad (4.10)$$

$(t_0 \leq s \leq t \leq T)$ and so

$$P(t \mid s) = \Phi(t, s)P(s)\Phi(t, s)' + \int_s^t \Phi(t, \tau)B(\tau)B(\tau)'\Phi(t, \tau)'\, d\tau \qquad (4.11)$$

Let $\tilde{\mathscr{Y}}_t$ be an arbitrary sub-σ-algebra of $\mathscr{Y}_t$, generated by a finite number of random variables $y(s_k)$, $t_0 \leq s_k \leq t$. By (4.5b) and (4.10), the joint distribution of the random variables $x(t), y(s_k)$ $(k = 1, 2, \ldots)$ is Gaussian. Hence $\text{cov}\{x(t) \mid \tilde{\mathscr{Y}}_t\}$ is a nonrandom matrix, and (4.4) and (4.11) provide the uniform bound

$$\text{cov}\{x(t) \mid \tilde{\mathscr{Y}}_t\} \leq \text{cov}\{x(t)\} = P(t \mid t_0) \leq \text{const}, \qquad t_0 \leq t \leq T, \quad (4.12)$$

independently of $\tilde{\mathscr{Y}}_t$. Similarly

$$\mathscr{E}\{|\mathscr{E}\{x(t) \mid \tilde{\mathscr{Y}}_t\}|^2\} \leq \mathscr{E}\{\mathscr{E}\{|x(t)|^2 \mid \tilde{\mathscr{Y}}_t\}\} = \mathscr{E}\{|x(t)|^2\} \leq \text{const},$$
$$t_0 \leq t \leq T. \quad (4.13)$$

Our procedure will be to compute a discrete-time approximation to $P(t)$ and from this show that $P(t)$ satisfies a certain differential equation. To justify the method we need some preliminary results.

Lemma 4.1. *Let* $\{y(t): t_0 \leq t \leq T\}$ *be a stochastically continuous[2] process in* R^m; *let* $S = \{s_n\}$ *be a denumerable set everywhere dense in* $[t_0, T]$; *and let*

$$\hat{\mathscr{Y}}_t = \sigma\{y(s): s \in [t_0, t] \cap S\}.$$

If f is a random variable such that $\mathscr{E}\{|f|\} < \infty$ *then*

$$\mathscr{E}\{f \mid \mathscr{Y}_t\} = \mathscr{E}\{f \mid \hat{\mathscr{Y}}_t\},$$

for every $t \in [t_0, T]$ *with probability* 1.

---

[2] A process $\{y(t)\}$ is *stochastically continuous* if $\text{P}\lim y(s) = y(t)$ $(s \to t)$ for each t.

Proof. Since $\mathcal{E}\{f \mid \hat{\mathcal{Y}}_t\}$ is $\mathcal{Y}_t$-measurable it is enough to check that

$$\int_\Lambda \mathcal{E}\{f \mid \hat{\mathcal{Y}}_t\} \mathbf{P}\,(d\omega) = \int_\Lambda f(\omega) \mathbf{P}\,(d\omega), \qquad (4.14)$$

for all sets Λ of the form

$$\Lambda = \{\omega\colon y(t_1\ \omega) \in \mathbf{B}_1, \ldots, y(t_r, \omega) \in \mathbf{B}_r\},$$

where $t_0 \le t_i \le t$ and $\mathbf{B}_i \in \mathcal{B}(R^m)$ $(i = 1, \ldots, r)$ (cf. the discussion by Doob [10, Chapter 1, pp. 19–20]). Clearly (4.14) is true if $\Lambda \in \hat{\mathcal{Y}}_t$. If $t' \in [t_0, t]$ is arbitrary there is a sequence $s_n \in S$ such that $s_n \to t'$ and hence by stochastic continuity,

$$\mathbf{P} \lim y(s_n) = y(t'), \qquad n \to \infty.$$

As the limit in probability of a sequence of $\hat{\mathcal{Y}}_t$-measurable functions, $y(t')$ is also $\hat{\mathcal{Y}}_t$-measurable. Hence $\Lambda \in \hat{\mathcal{Y}}_t$ for arbitrary choices of the t_i, and (4.14) follows.

Lemma 4.2. *Let* $\{t_j^{(n)}\colon j = 0, 1, \ldots, N_n\}$, $n = 1, 2, \ldots$, *be an increasing sequence of partitions of* $[t_0, T]$ *such that*

$$t_0 = t_0^{(n)} < t_1^{(n)} < \cdots < t_{N_n}^{(n)} = T,$$

and

$$\lim_{n \to \infty}\ \max_{1 \le j \le N_n} (t_j^{(n)} - t_{j-1}^{(n)}) = 0.$$

Let $\{y(t)\colon t_0 \le t \le T\}$, $\mathcal{Y}_t$, *and* f *be as in Lemma* 4.1; *and for* $t_0 \le t \le T$ *let*

$$\mathcal{Y}_t^{(n)} = \sigma\{y(t_j^{(n)})\colon 0 \le j \le N_n, t_j^{(n)} \le t\}.$$

Then

$$\mathcal{E}\{f \mid \mathcal{Y}_t^{(n)}\} \to \mathcal{E}\{f \mid \mathcal{Y}_t\}, \qquad n \to \infty,$$

with probability 1, *and in the mean.*

Proof. Since $\mathcal{E}\{|f|\} < \infty$ the sequence $\mathcal{E}\{f \mid \mathcal{Y}_t^{(n)}\}$ $(n = 1, 2, \ldots)$ is a martingale and, by known convergence theorems (Doob [10, Theorem 4.1, p. 319 and Theorem 4.3, p. 331]), the limit

$$\lim_{n \to \infty} \mathcal{E}\{f \mid \mathcal{Y}_t^{(n)}\} = \mathcal{E}\{f \mid \hat{\mathcal{Y}}_t\},$$

exists with probability 1 and in the mean, where

$$\hat{\mathcal{Y}}_t = \sigma\{y(t_j^{(n)})\colon 0 \le j \le N_n, t_j^{(n)} \le t; n = 1, 2, \ldots\}.$$

Since the set

$$\bigcup_{n=1}^{\infty} \{t_j^{(n)} : 0 \le j \le N_n, t_j^{(n)} \le t\},$$

is dense in $[t_0, t]$, it follows by Lemma 4.1 that

$$\mathscr{E}\{f \mid \hat{\mathscr{Y}}_t\} = \mathscr{E}\{f \mid \mathscr{Y}_t\}.$$

The proof is complete.

In the following we retain the notation of Lemmas 4.1 and 4.2. Returning to the filtering problem, let

$$\hat{x}_n(t) = \mathscr{E}\{x(t) \mid \mathscr{Y}_t^{(n)}\},$$
$$P_n(t) = \text{cov}\{x(t) \mid \mathscr{Y}_t^{(n)}\}. \qquad (4.15)$$

By (4.12) and (4.13), the quantities $\mathscr{E}\{|\hat{x}_n(t)|^2\}$ and $|P_n(t)|$ are bounded on $[t_0, T]$ uniformly in n; since the $y(t)$ process is clearly stochastically continuous, Lemma 4.2 implies

$$\hat{x}_n(t) \to \hat{x}(t), \qquad n \to \infty,$$

with probability 1 and in mean square; hence also

$$P_n(t) \to P(t), \qquad n \to \infty.$$

C. Derivation of the Filter Equations

Let n be fixed. For $0 \le j \le N_n$ we write $P_j, x_j, y_j, \mathscr{Y}_j, t_j$ for $P_n(t_j^{(n)})$, $x(t_j^{(n)}), y(t_j^{(n)}), \mathscr{Y}_{t_j}^{(n)}, t_j^{(n)}$, respectively. Let $\delta_n = (T - t_0)/N_n$ and $t_j^{(n)} = t_0 + j\delta_n, 1 \le j \le N_n$. Writing $\delta_n = \delta$, we have

$$\Phi(t_{j+1}, t_j) = I + \delta A(t_j) + O(\delta^2), \qquad (\delta \downarrow 0). \qquad (4.16)$$

To compute the difference $P_{j+1} - P_j$, we form the *a priori* estimate (conditional expectation) of x_{j+1} given the y_i for $i \le j$, and then use (4.4) to calculate the *a posteriori* covariance of x_{j+1}. Write

$$\hat{x}(t \mid j) = \mathscr{E}\{x(t) \mid \mathscr{Y}_j\},$$
$$P_{j+1}(j) = \text{cov}\{x_{j+1} \mid \mathscr{Y}_j\}.$$

Then from (4.10) and (4.11)

$$\hat{x}(t \mid j) = \Phi(t, t_j)\hat{x}_j, \qquad (4.17)$$

and

$$P_{j+1}(j) = \Phi(t_{j+1}, t_j)P_j\Phi(t_{j+1}, t_j)'$$
$$+ \int_{t_j}^{t_{j+1}} \Phi(t_{j+1}, t)B(t)B(t)'\Phi(t_{j+1}, t)' \, dt. \qquad (4.18)$$

The matrix $P_{j+1}(j)$ will play the role of the matrix P in (4.1). To compute the corresponding matrices Q and R we note first that

$$y_{j+1} = y_j + \int_{t_j}^{t_{j+1}} C(t)x(t) \, dt + \int_{t_j}^{t_{j+1}} D(t) \, dw(t),$$

whence

$$\mathscr{E}\{y_{j+1} \mid \mathscr{Y}_j\} = y_j + \int_{t_j}^{t_{j+1}} C(t)\hat{x}(t \mid j) \, dt$$

$$= y_j + \int_{t_j}^{t_{j+1}} C(t)\Phi(t, t_j)\hat{x}_j \, dt$$

$$= y_j + \left(\int_{t_j}^{t_{j+1}} C(t) \, dt\right)\hat{x}_j + \delta^2\theta_j \qquad (4.19)$$

where $\mathscr{E}\{|\theta_j|^2\} = 0(1)$ as $\delta \downarrow 0$, uniformly in j. To evaluate R, write

$$\Phi(t, t_j)(x_j - \hat{x}_j) = \Delta_j(t),$$

$$\int_{t_j}^{t} \Phi(t, s)B(s) \, dw(s) = \xi_j(t),$$

$$\int_{t_j}^{t_{j+1}} D(t) \, dw(t) = \eta_j.$$

Then $x(t) = \Phi(t, t_j)x_j + \xi_j(t)$, and

$$y_{j+1} - y_j = \int_{t_j}^{t_{j+1}} C(t)[\Phi(t, t_j)x_j + \xi(t)] \, dt + \eta_j.$$

Hence

$$R = \text{cov}\{y_{j+1} \mid \mathscr{Y}_j\} = \text{cov}\left\{\int_{t_j}^{t_{j+1}} C(t)[\Delta_j(t) + \xi_j(t)] \, dt + \eta_j \mid \mathscr{Y}_j\right\}.$$

Since $\xi_j(t)$ $(t \geq t_j)$ and η_j are independent of $\mathscr{Y}_j$,

$$R = \text{cov}\left\{\int_{t_j}^{t_{j+1}} C(t)\Delta_j(t) \, dt \mid \mathscr{Y}_j\right\} + \text{cov}\left\{\eta_j + \int_{t_j}^{t_{j+1}} C(t)\xi_j(t) \, dt\right\}$$

$$= \int_{t_j}^{t_{j+1}} D(t)D(t)' \, dt + O(\delta^2), \qquad (\delta \downarrow 0). \qquad (4.20)$$

Because $DD' \geq \varepsilon I$, (4.20) shows that δR^{-1} exists and is bounded as $\delta \downarrow 0$; i.e.,

$$\delta R^{-1} = \delta \left[\int_{t_j}^{t_{j+1}} D(t)D(t') \, dt \right]^{-1} + O(\delta) \qquad (\delta \downarrow 0). \qquad (4.21)$$

Similarly

$$Q = \text{cov}\{x_{j+1}, y_{j+1} \mid \mathscr{Y}_j\}$$

$$= \mathscr{E}\left\{ [x_{j+1} - \Phi(t_{j+1}, t_j)\hat{x}_j] \right.$$

$$\left. \cdot \left[\int_{t_j}^{t_{j+1}} C(t)\{x(t) - \Phi(t, t_j)\hat{x}_j\} \, dt + \int_{t_j}^{t_{j+1}} D(t) \, dw(t) \right]' \mid \mathscr{Y}_j \right\}$$

$$= \mathscr{E}\left\{ [\Delta_j(t_{j+1}) + \xi_j(t_{j+1})] \cdot \left[\int_{t_j}^{t_{j+1}} C(t)\{\Delta_j(t) + \xi_j(t)\} \, dt + \eta_j \right]' \mid \mathscr{Y}_j \right\}$$

$$= \int_{t_j}^{t_{j+1}} [P_j C(t)' + B(t)D(t)'] \, dt + O(\delta^{3/2}), \qquad (\delta \downarrow 0). \qquad (4.22)$$

It is now possible to compute

$$P_{j+1} = P_{j+1}(j) - QR^{-1}Q, \qquad (4.23)$$

By (4.16), (4.18), (4.20), and (4.21) we have

$$P_{j+1} - P_j = \int_{t_j}^{t_{j+1}} [A(t)P_j + P_j A(t)' + B(t)B(t)'] \, dt$$

$$- \left(\int_{t_j}^{t_{j+1}} [P_j C(t)' + B(t)D(t)'] \, dt \right)\left(\int_{t_j}^{t_{j+1}} D(t)D(t)' \, dt \right)^{-1}$$

$$\cdot \left(\int_{t_j}^{t_{j+1}} [P_j C(t)' + B(t)D(t)'] \, dt \right)' + O(\delta^{3/2})$$

$$= \int_{t_j}^{t_{j+1}} \{A(t)P_j + P_j A(t)' + B(t)B(t)'$$

$$+ [P_j C(t)' + B(t)D(t)'][D(t)D(t)']^{-1}$$

$$\times [P_j C(t)' + B(t)D(t)']'\} \, dt + O(\delta^{3/2}) \qquad (4.24)$$

where $O(\cdot)$ is uniform in j. We now indicate explicitly the dependence of P_j, etc., on n. For $t_j^{(n)} \leq t < t_{j+1}^{(n)}$, $P_n(t)$ is given by (4.11) with $P(s)$ replaced by P_j. This implies that $P_j(=P_n(t_j^{(n)}))$ may be replaced in

(4.24) by $P_n(t)$, with error of $O(\delta_n)$ as $\delta_n \downarrow 0$. With this replacement (4.24) yields

$$P_n(T) - P(t_0) = \sum_{j=1}^{N_n} [P_n(t_j^{(n)}) - P_n(t_{j-1}^{(n)})]$$

$$= \int_{t_0}^{T} [AP_n + P_n A' + BB'$$

$$+ (P_n C' + BD')(DD')^{-1}(P_n C' + BD')'] \, dt$$

$$+ \text{const. } N_n \delta_n^2 + N_n O(\delta_n^{3/2}). \qquad (4.25)$$

Denote the integral in (4.25) by $J(P_n)$. With the partitions $\{t_j^{(n)}\}$ chosen such that $\delta_n = N_n^{-1}(T - t_0)$, there follows

$$P_n(T) - P(t_0) - J(P_n) \to 0, \qquad n \to \infty. \qquad (4.26)$$

Now (cf. (4.11)) the $P_n(\cdot)$ are right-continuous functions, in general having jumps at the points $t_j^{(n)}$; we have seen that $P(t) = \lim P_n(t)$ $(n \to \infty)$ exists, and that the $P_n(\cdot)$ are uniformly bounded. By dominated convergence, $J(P_n) \to J(P)$ $(n \to \infty)$, and there follows

$$P(T) = P(t_0) + J(P).$$

The above argument applies on every subinterval $t_0 \le t' \le t \le T$; and from this we conclude that $P(\cdot)$ satisfies the *matrix Riccati equation*

$$dP/dt = AP + PA' + BB' + (PC' + BD')(DD')^{-1}(PC' + BD')'$$

$$t_0 \le t \le T, \qquad (4.27)$$

$$P(t_0) = \text{cov}\{x_0\}.$$

We now show that the estimate $\hat{x}(t)$ can be represented as the solution of a stochastic differential equation. In the simplified notation used before,

$$\hat{x}_{j+1} = x(t_{j+1}|j) + QR^{-1}[y_{j+1} - \mathscr{E}\{y_{j+1}|\mathscr{Y}_j\}].$$

Evaluating the quantities on the right from (4.17), (4.19), (4.21), (4.22), there results

$$\hat{x}_{j+1} = \Phi(t_{j+1}, t_j)\hat{x}_j$$

$$+ \left(\int_{t_j}^{t_{j+1}} [P_j C(t)' + B(t)D(t)'] \, dt\right)\left(\int_{t_j}^{t_{j+1}} D(t)D(t)' \, dt\right)^{-1}$$

$$\cdot \left[y_{j+1} - y_j - \int_{t_j}^{t_{j+1}} C(t)\hat{x}_j \, dt\right] + \delta^2 \hat{\xi}_j, \qquad (4.28)$$

where $\hat{\xi}_j$ denotes a random variable such that $\mathscr{E}\{|\hat{\xi}_j|^2\} = O(1)$ as $\delta \downarrow 0$, uniformly in j. Replacing P_j by $P_n(t)$, using the fact that $|P_n(t) - P_j| < \text{const.} (t - t_j)$, $t_j \le t < t_{j+1}$, and recalling (4.17), we can write (4.28) as

$$\hat{x}_n(t_{j+1}) = \hat{x}_n(t_j) + \int_{t_j}^{t_{j+1}} A(t)\hat{x}_n(t)\, dt$$

$$+ \int_{t_j}^{t_{j+1}} K_n(t)[dy(t) - C(t)\hat{x}_n(t)\, dt] + \delta_n^2 \hat{\xi}_{jn} \quad (4.29)$$

where $\mathscr{E}\{|\hat{\xi}_{jn}|^2\} < M$ is uniformly bounded in j, n as $n \to \infty$, $t_j = t_j^{(n)}$, etc., and

$$K_n(t) = [P_n(t)C(t)' + B(t)D(t)'][D(t)D(t)']^{-1}.$$

Define the processes $\{\tilde{z}_n(t): t_0 \le t \le T\}$ $(n = 1, 2, \ldots)$ by

$$\tilde{z}_n(t_{j+1}) = \tilde{z}_n(t_j) + \int_{t_j}^{t_{j+1}} A(t)\tilde{z}_n(t)\, dt$$

$$+ \int_{t_j}^{t_{j+1}} K_n(t)[dy(t) - C(t)\tilde{z}_n(t)]\, dt, \quad (4.30)$$

$$j = 0, 1, \ldots, N_n - 1.$$

$$\tilde{z}_n(t) = \Phi(t, t_j)\tilde{z}_n(t_j), \quad t_j \le t < t_{j+1},$$

$$\tilde{z}_n(t_0) = \hat{x}(t_0).$$

Writing $\tilde{\Delta}_n(t) = \tilde{z}_n(t) - \hat{x}_n(t)$, there follows

$$|\tilde{\Delta}_n(t_{j+1})| \le |\tilde{\Delta}_n(t_j)| + \theta\delta_n |\tilde{\Delta}_n(t_j)| + \delta_n^2 |\hat{\xi}_{jn}|$$

where θ is a constant independent of j, n. Hence

$$\mathscr{E}\{|\tilde{\Delta}_n(t_j)|^2\} \le \left\{\frac{(1 + \theta\delta_n)^{N_n} - 1}{\theta\delta_n}\right\}^2 \delta_n^4 M$$

With $\delta_n = N_n^{-1}(T - t_0)$ there results

$$\mathscr{E}\{|\tilde{\Delta}_n(t_j^{(n)})|^2\} \to 0, \quad n \to \infty.$$

Hence for $t_0 \le t \le T$,

$$\mathscr{E}\{|\tilde{z}_n(t) - \hat{x}_n(t)|^2\} \to 0, \quad n \to \infty,$$

and therefore

$$\mathscr{E}\{|\tilde{z}_n(t) - \hat{x}(t)|^2\} \to 0, \quad n \to \infty. \quad (4.31)$$

Now define $\{z_n(t): t_0 \le t \le T\}$ as the solution of

$$dz_n = A(t)z_n(t)\, dt + K_n(t)[dy(t) - C(t)z_n(t)], \qquad t_0 \le t \le T,$$
$$z_n(t_0) = \hat{x}(t_0); \qquad\qquad\qquad\qquad\qquad\qquad\qquad\quad (4.32)$$

and write $\Delta_n(t) = \tilde{z}_n(t) - z_n(t)$. Then

$$\Delta_n(t_{j+1}) = \Delta_n(t_j) + \int_{t_j}^{t_{j+1}} [A(t) - K_n(t)C(t)]\Delta_n(t)\, dt,$$

and for $t_j \le t < t_{j+1}$,

$$\Delta_n(t) = \Phi(t, t_j)\tilde{z}_n(t_j) - z_n(t_j) - \int_{t_j}^{t} K_n(s)[dy(s) - C(s)z_n(s)]\, ds \quad (4.33)$$

Recalling $dy = Cx\, dt + D\, dw$, there results, for $j = 0, 1, \ldots, N_n - 1$,

$$\Delta_n(t_{j+1}) - \Delta_n(t_j) - \theta_{jn}\delta_n^2$$
$$= - \int_{t_j}^{t_{j+1}} [A(t) - K_n(t)C(t)]\left\{\int_{t_j}^{t} K_n(s)D(s)\, dw(s)\right\} dt, \quad (4.34)$$

where $\mathscr{E}\{|\theta_{jn}|^2\} = O(1)$ uniformly in j, n. The random variables on the right side of (4.34) are independent and have variance of order $O(\delta_n^3)$. Therefore, with $\delta_n = N_n^{-1}(T - t_0)$,

$$\mathscr{E}\left\{\left|\sum_{j=0}^{k-1}[\Delta_n(t_{j+1}) - \Delta_n(t_j) - \theta_{jn}\delta_n^2]\right|^2\right\} = O(\delta_n^2)$$

or

$$\mathscr{E}\{|\Delta_n(t_k) - \tilde{\theta}_{kn}\delta_n|^2\} = O(\delta_n^2),$$

where $\mathscr{E}\{|\tilde{\theta}_{kn}|^2\} = O(1)$ uniformly in j, n. Hence

$$\mathscr{E}\{|\Delta_n(t_k)|^2\} \to 0, \qquad n \to \infty,$$

for $1 \le k \le N_n$. It follows from (4.32) that

$$\mathscr{E}\{|\tilde{z}_n(t) - z_n(t)|^2\} \to 0, \qquad n \to \infty, \qquad\qquad (4.35)$$

for $t_0 \le t \le T$.

Finally, let $\{z(t): t_0 \le t \le T\}$ be the solution of

$$dz(t) = A(t)z(t) + K(t)[dy(t) - C(t)z(t)], \qquad t_0 \le t \le T,$$
$$z(t_0) = \hat{x}(t_0). \qquad\qquad\qquad\qquad\qquad\qquad\qquad\quad (4.36)$$

To complete the discussion we observe that (4.25)–(4.27), together with Gronwall's inequality, imply that $P_n \to P$ uniformly on $[t_0, T]$, and therefore $K(t) = \lim K_n(t)(n \to \infty)$ exists uniformly. Using this fact it is easy to see from (4.32) and (4.36) that

$$\mathscr{E}\{|z_n(t) - z(t)|^2\} \to 0, \qquad n \to \infty, \qquad (4.37)$$

for $t_0 \leq t \leq T$. By (4.31), (4.35), and (4.37) we can now conclude that $\hat{x}(t)$ is the solution of (4.36).

To summarize results, we have

Theorem 4.1. *Let* $\{x(t), y(t); t_0 \leq t \leq T\}$ *be the Gaussian processes determined by* (4.5) *and* (4.6). *If*

$$\hat{x}(t) = \mathscr{E}\{x(t) \mid \mathscr{Y}_t\}, \qquad t_0 \leq t \leq T,$$

then $\hat{x}(t)$ *is characterized as the solution of the stochastic differential equation* (4.36), *where*

$$\hat{x}(t_0) = \mathscr{E}\{x_0\}.$$

In (4.36),

$$K(t) = [P(t)C(t)' + B(t)D(t)'][D(t)D(t)']^{-1}.$$

Furthermore

$$P(t) = \text{cov}\{x(t) \mid \mathscr{Y}_t\}, \qquad t_0 \leq t \leq T,$$

and $P(t)$ *satisfies the matrix Riccati equation* (4.27).

Equations (4.27) and (4.36) were first derived, in a different (and quite formal) way, by Kalman and Bucy [36]; the present discussion is based on Wonham [79]. The equations prescribe the dynamic structure of a filter (analog device) for generating the estimate $x(t)$, in real time, from noisy data given by the formal equation $\dot{y} = Cx + D\dot{w}$. To simulate the filter, one first solves (4.27) and stores $P(t)$ for $t_0 \leq t \leq T$. The filter itself is specified by (4.36) or actually, for simulation purposes, by the formal version

$$d\hat{x}/dt = A\hat{x} + (PC' + BD')(DD')^{-1}(\dot{y} - C\hat{x}), \qquad t_0 \leq t \leq T. \quad (4.38)$$

Thus (4.38) is solved in "real time" as the noisy data $\dot{y}$ becomes available.

In engineering practice (4.27) and (4.38) are usually solved in discrete approximation. Derivation of the discrete-time counterparts of (4.27) and (4.38) requires merely the application of (4.2) and (4.3) to the difference equation of a discrete-time model.

The existence, uniqueness, and asymptotic behavior of the solution of the Riccati equation (4.27) have been discussed in detail elsewhere (Kalman [35]; Wonham [81]). It is known that (4.27) has a unique solution on every finite interval $[t_0, T]$. If the parameter matrices are all constant then, under suitable observability conditions (see, e.g., Wonham [81]), $P_\infty = \lim P(t) \, (t \to \infty)$ exists, and the matrix

$$A - (P_\infty C' + BD')(DD')^{-1}C$$

is stable. The resulting stable, time-invariant filter is precisely the Wiener filter for stationary least-squares smoothing.

V. The Separation Theorem

A. INTRODUCTION

In this section the filtering theory of Section IV is applied to the problem of optimal control of a system with linear dynamics, when the state information available to the controller is perturbed by additive Gaussian white noise. A proof is given of the so-called "separation theorem" of stochastic control, according to which the problem of filtering and that of control proper can be treated, to some extent, independently. This structural property of the optimal system holds whether or not the cost functional is quadratic, and whether or not the optimal control happens to be linear in the system state or its expectation. In general, the optimal control depends on the parameters of the channel noise process. The separation property means, however, that channel noise plays qualitatively the same role as dynamic disturbances in determination of the feedback law.

This result can be regarded as an existence theorem which, in effect, vindicates the approach by dynamic programming to a fairly general class of problems. A special result of this type for the standard, linear stochastic regulator problem, is well-known, and is the original "separation theorem": Joseph and Tou [32], Potter [57].

For discrete-time systems the general result can be proved by relatively straightforward application of dynamic programming (Striebel [66]). In this section, attention is confined to continuous systems. The method is again dynamic programming, with appeal to the Itô–Nisio–Fleming theory of functional stochastic differential equations (Fleming and Nisio [18]), Kalman's filter (Kalman [35]), and an existence theorem for parabolic equations due to Ladizhenskaya *et al.* [47]. Since the aim is to clarify the principles involved, simplifying technical hypotheses are imposed which are not always met in practice. Nevertheless the result obtained covers a wide enough class of situations to be of practical interest.

The discussion below closely follows Wonham [80].

B. STATEMENT OF THE PROBLEM

The system to be controlled is described by linear stochastic differential equations:

$$dx(t) = A(t)x(t) \, dt + b[t, u(t)] \, dt + C(t) \, dw_1(t), \qquad 0 \le t \le T,$$
$$x(0) = x_0, \tag{5.1}$$

$$dy(t) = F(t)x(t) \, dt + G(t) \, dw_2(t), \qquad 0 \le t \le T,$$
$$y(0) = 0. \tag{5.2}$$

Here and below all vectors and matrices have real-valued elements. The vector $x \in R^n$ is dynamic state; u is the control vector taking values in a convex compact subset $U \subset R^m$; $y \in R^n$ is channel output; and w_1, w_2 are independent, standard Wiener processes in R^{d_1}, R^{d_2}, respectively.

In practical terms, the problem is to control $x(\cdot)$ in such a way as to minimize a real-valued functional

$$J[u] = \mathscr{E}\left\{\int_0^T L[t, x(t), u(t)] \, dt\right\} \tag{5.3}$$

Control is based on the (*a priori*) distribution of x_0 and on information provided by the channel output $y(\cdot)$. Since the controller is not clairvoyant, $u(t)$ must be assumed to depend only on the $y(s)$ for $0 \le s \le t$. To express this *nonanticipative* dependence we introduce, following Fleming and Nisio [18], a suitable class of control functionals. Let $\mathscr{C}$ denote the class of functions $f(t)$ continuous on $[0, T]$ with values in R^n; and write, for the *past of f at time t*,

$$(\pi_t f)(s) = \begin{cases} f(s), & 0 \le s \le t \\ f(t), & t \le s \le T \end{cases} \tag{5.4}$$

Clearly $\pi_t f \in \mathscr{C}$ if $f \in \mathscr{C}$. Let $\|\cdot\|$ denote sup norm in $\mathscr{C}$ and let

$$\psi: [0, T] \times \mathscr{C} \to U,$$

be a mapping with the properties: $\psi(t, f)$ is Hölder continuous in t for each $f \in \mathscr{C}$ and satisfies a uniform Lipschitz condition[3]

$$|\psi(t, f) - \psi(t, g)| < c_1 \|f - g\|, \qquad (5.5)$$

$(t \in [0, T]; f, g \in \mathscr{C})$. Let Ψ denote the class of functionals ψ. We call the control $u(\cdot)$ *admissible*, and write $u \in \mathscr{U}$, if

$$u(t) = \psi(t, \pi_t y), \qquad 0 \le t \le T,$$

for some $\psi \in \Psi$. *The problem is to find $u^0 \in \mathscr{U}$ such that*

$$J[u^0] = \min\{J[u] : u \in \mathscr{U}\}.$$

The corresponding functional ψ^0 is *optimal*. It will be verified later that $J[u]$ is well-defined.

The separation theorem states that an optimal control exists in a subclass $\hat{\mathscr{U}}$ of controls which depend only on the expected value of the current state given the past of y. More precisely, let

$$\mathscr{Y}_t = \sigma\{y(s), 0 \le s \le t\},$$
$$\hat{x}(t) = \mathscr{E}\{x(t) \mid \mathscr{Y}_t\}.$$

Write $\hat{\Psi}$ for the class of functions

$$\hat{\psi}: [0, T] \times R^n \to U,$$

such that

$$|\hat{\psi}(t, \xi) - \hat{\psi}(s, \xi)| + |\hat{\psi}(t, \xi) - \hat{\psi}(t, \eta)| \le c_2(R)|t - s|^\alpha + c_3|\xi - \eta|, \quad (5.6)$$

in every domain $0 \le s, t \le T, |\xi| < R, |\eta| < R$, where c_3, and $\alpha \in (0, \frac{1}{2})$, are independent of R. We write $u \in \hat{\mathscr{U}}$ if

$$u(t) = \hat{\psi}[t, \hat{x}(t)], \qquad t \in [0, T],$$

for some $\hat{\psi} \in \hat{\Psi}$. It will be shown later that $\hat{\mathscr{U}} \subset \mathscr{U}$.

The following additional assumptions will be made. We write uhc(α) for "uniformly Hölder continuous (exponent α)," and ulc for "uniformly Lipschitz continuous," where the uniformity is to hold over the whole range of the relevant arguments, unless otherwise stated.

A.1 The matrices A, C are uhc(α) in t, and F, G are continuously differentiable in $[0, T]$.

A.2 $G(t)G(t)' \ge c_4 I$, $t \in [0, T]$.

A.3 $|\det[F(t)]| \ge c_5$, $t \in [0, T]$.

---

[3] Here and below, $c_1 \, c_2, \ldots$ denote positive constants.

A.4 b, b_u, b_{uu} are continuous on $[0, T] \times U$ and b, b_u are uhc(α) in t.

A.5 L and L_u are bounded uhc(α) in t, and ulc in x. L_{uu} is bounded and uniformly continuous on $[0, T] \times R^n \times U$.

A.6 $[b(t, u)'p + L(t, x, u)]_{uu} \geq c_6 I,$ for all

$$(t, x, u, p) \in [0, T] \times R^n \times U \times \{p : |p| \leq \pi\},$$

where π is defined by (5.41), below.

A.7 x_0 is a Gaussian random variable independent of the processes $w_1(t)$, $w_2(t)$, and with positive definite covariance matrix Q_0.

The foregoing restrictions are mainly technical. A.3 would rarely be met in practice, where typically dim $y <$ dim x; the condition is needed below to guarantee that a certain elliptic operator be nondegenerate. A square nonsingular matrix F could be constructed artificially, if necessary, by adjoining to the channel equation (5.2) a suitable term of form

$$d\tilde{y} = \varepsilon \tilde{F} \tilde{x} \, dt + \tilde{G} \, d\tilde{w}_2.$$

If $\varepsilon > 0$ is sufficiently small then, from a practical viewpoint, the components $\tilde{y}$ of the observation vector contribute negligible information to the controller. However, details of such an approximation have yet to be worked out.

The number π in A.6 is an *a priori* bound on the space-derivative of the solution of Bellman's equation. In the special but important case where $b(t, u)$ is linear in u, the estimate π is not required, and A.6 can be replaced by A.6'

$$L_{uu}(t, x, u) \geq c_6 I, \qquad (t, x, u) \in [0, T] \times R^n \times U.$$

The crucial assumptions for Theorem 5.1 are that the basic dynamic equations have the form (5.1), (5.2); that the (formal) perturbations dw_i/dt ($i = 1, 2$) be "white Gaussian noise"; that x_0 be Gaussian and independent of the w_i; and that $J[u]$ be a functional additive in t.

Theorem 5.1 (Separation Theorem). *Subject to the assumptions stated, an admissible optimal control exists, in the subclass $\hat{\mathscr{U}} \subset \mathscr{U}$; that is,*

$$u^0(t) = \hat{\psi}^0[t, \hat{x}(t)], \qquad t \in [0, T],$$

for some $\hat{\psi}^0 \in \hat{\Psi}$.

An optimal feedback law $\hat{\psi}^0$ is given by (5.42), below.

The theorem will be proved in several steps. In Section V.C we verify that the solution of (5.1), (5.2) is well-defined. Kalman's equations for $\hat{x}(t)$ are introduced in Section V.D, and it is shown that $\hat{x}(t)$ is a diffusion process when $u = \hat{\psi} \in \hat{\Psi}$. In Section V.E we prove that Bellman's equation provides a sufficient condition for optimality, and in Section V.F, that an optimal control exists. The standard linear regulator problem lies outside the scope of Theorem 5.1, and is discussed separately in Section V.G.

C. Solution of (5.1) and (5.2)

As usual, (5.1) and (5.2) are to be interpreted as stochastic integral equations, with integrals defined in the sense of Itô. Let $\psi \in \Psi$ and $f, g \in \mathscr{C}$. From (5.4), (5.5) and A.4 it follows that

$$|b[t, \psi(t, \pi_t f)] - b[t, \psi(t, \pi_t g)]|$$

$$\leq \max_{\substack{t \in [0, T] \\ u \in U}} |b_u(t, u)| \cdot c_7 \|\pi_t f - \pi_t g\|$$

$$\leq c_8 \|f - g\|. \tag{5.7}$$

Clearly $b[t, \psi(t, \pi_t f)]$ is bounded on $[0, T] \times \mathscr{C}$. It follows by Theorem 1 of Fleming and Nisio [18] that the system (5.1), (5.2) has exactly one continuous solution $\{x(t), y(t): 0 \leq t \leq T\}$ with bounded second moment. Furthermore, this solution has the property that $x(t), y(t)$ are measurable relative to $\sigma\{x_0, w_1(s), w_2(s), 0 \leq s \leq t\}$ and are independent of $\sigma\{w_i(t'') - w_i(t'), t \leq t' \leq t'' \leq T, i = 1, 2\}$. By continuity and (Doob [10, p. 60, Theorem 2.5]), $\{x(t), y(t)\}$ is a measurable process.

D. Representation of $\hat{x}(t)$ as a Diffusion Process

Suppose $b \equiv 0$. The problem of representing $\hat{x}(t)$ as the solution of a stochastic differential equation was solved in Section IV, and so we need justify only the extension for $b \not\equiv 0$. Let $u(\cdot)$ be admissible and write

$$\beta(t) = b[t, u(t)] = b[t, \psi(t, \pi_t y)]. \tag{5.8}$$

Observe first that the random variable $\beta(t)$ is $\mathscr{Y}_t$-measurable. Indeed if S is open in R^m then $\tilde{S} = \{f: b[t, \psi(t, f)] \in S\}$ is open in $\mathscr{C}$; and using the fact that $\mathscr{C}$ is a separable metric space, we see that

$$\{\omega: \pi_t y(\cdot, \omega) \in \tilde{S}\} \in \mathscr{Y}_t.$$

The assertion follows by extension to the Borel sets of R^m. Next, let

$$x(t) = \tilde{x}(t) + x^*(t),$$

where $\tilde{x}(t)$ is the diffusion process determined by

$$d\tilde{x}(t) = A(t)\tilde{x}(t)\, dt + C(t)\, dw_1(t), \qquad t \in [0, T],$$
$$\tilde{x}(0) = x_0; \tag{5.9}$$

and $x^*(t)$ is defined by

$$dx^*(t)/dt = A(t)x^*(t) + \beta(t), \qquad t \in [0, T],$$
$$x^*(0) = 0. \tag{5.10}$$

Since $x^*(t)$ is $\mathcal{Y}_t$-measurable there follows

$$\hat{x}(t) = \mathscr{E}\{\tilde{x}(t) \mid \mathcal{Y}_t\} + x^*(t). \tag{5.11}$$

Now define a process $\tilde{y}(t)$ according to

$$d\tilde{y}(t) \equiv dy(t) - F(t)x^*(t)\, dt \tag{5.12a}$$
$$d\tilde{y}(t) = F(t)\tilde{x}(t)\, dt + G(t)\, dw_2(t), \qquad t \in [0, T], \tag{5.12b}$$
$$\tilde{y}(0) = 0;$$

and let

$$\tilde{\mathcal{Y}}(t) = \sigma\{\tilde{y}(s), 0 \le s \le t\}.$$

By (5.10), $x^*(t)$ is $\mathcal{Y}_t$-measurable; then, by (5.12), $\tilde{y}(t)$ is $\mathcal{Y}_t$-measurable. It will be shown that $y(t)$ is $\tilde{\mathcal{Y}}_t$-measurable, and thus that $\mathcal{Y}_t = \tilde{\mathcal{Y}}_t$. In view of (5.8) and (5.10), (5.12) can be written

$$y(t) = \tilde{y}(t) + \int_0^t F(s)x^*(s)\, ds$$
$$= \tilde{y}(t) + \phi(t, \pi_t y), \tag{5.13}$$

where $\phi \in \Psi$. Since $\phi(t, f)$ is bounded on $[0, T] \times \mathscr{C}$, and

$$|\phi(t, f) - \phi(t, g)| \le c_9 t \|f - g\|,$$

the functional equation (5.13) can be solved by successive approximations uniquely for y in $\mathscr{C}$. Setting $y^{(0)}(t) \equiv 0$ and

$$y^{(v)}(t) = \tilde{y}(t) + \phi(t, \pi_t y^{(v-1)}),$$

($v = 1, 2, \ldots$) we see that $y^{(v)}(t)$ is $\tilde{\mathcal{Y}}_t$-measurable for each v, and the conclusion follows.

We now have, from (5.11),

$$\hat{x}(t) = \bar{x}(t) + x^*(t), \tag{5.14}$$

where

$$\bar{x}(t) = \mathscr{E}\{\tilde{x}(t) \mid \tilde{\mathscr{Y}}(t)\}.$$

It remains to compute $\bar{x}(t)$. Equations (5.9) and (5.12b) have the form of (5.1) and (5.2) with $b \equiv 0$, and Theorem 3.4 applies. Introduce the conditional covariance matrix

$$Q(t) = \mathscr{E}\{[x(t) - \hat{x}(t)][x(t) - \hat{x}(t)]' \mid \mathscr{Y}_t\}$$
$$= \mathscr{E}\{[\tilde{x}(t) - \bar{x}(t)][\tilde{x}(t) - \bar{x}(t)]' \mid \tilde{\mathscr{Y}}_t\},$$

where the second equality holds because $x(t) = \tilde{x}(t) + x^*(t)$ and $\mathscr{Y}_t = \tilde{\mathscr{Y}}_t$. By Theorem 4.1, applied to (5.9) and (5.12b), $Q(t)$ is the unique solution of the Riccati equation

$$dQ/dt = AQ + QA' + CC' - QF'(GG')^{-1}FQ, \qquad t \in [0, T], \tag{5.15}$$
$$Q(0) = Q_0.$$

According to Theorem 4.1, $\bar{x}$ is determined by

$$d\bar{x} = A\bar{x}\, dt + QF'(GG')^{-1}(d\tilde{y} - F\bar{x}\, dt), \qquad t \in [0, T], \tag{5.16}$$

with initial condition

$$\bar{x}(0) = \mathscr{E}\{\tilde{x}(0) \mid \mathscr{Y}_0\} = \mathscr{E}\{\tilde{x}(0)\} = \mathscr{E}\{x_0\}.$$

Combining (5.9)–(5.12) and (5.14)–(5.16) there results finally

$$d\hat{x} = A\hat{x}\, dt + \beta(t)\, dt + QF'(GG')^{-1}(dy - F\hat{x}\, dt), \tag{5.17}$$
$$\hat{x}(0) = \mathscr{E}\{x_0\}.$$

Equation (5.17) exhibits the $\hat{x}(t)$ process as the solution of an equation "forced" by the channel output increments dy and by the control term β. It is possible—and for later purposes necessary—to replace the differential $dy - Fx\, dt$ by the suitably scaled differential of a Wiener process. This can be justified by the observation that linear least squares estimation is equivalent to an orthogonal projection of the estimated variable on the data; but we shall carry out the computation explicitly.

Let $e(t) = x(t) - \hat{x}(t)$ be the error process and define the *innovation process* $z(t)$ by

$$dz = dy - F\hat{x}\, dt$$
$$\quad = Fe\, dt + G\, dw_2, \qquad t \in [0, T], \tag{5.18}$$
$$z(0) = 0.$$

Evidently $z(t)$ is $\mathcal{Y}_t$-measurable. The relation

$$z(t_2) = z(t_1) + \int_{t_1}^{t_2} F(t)e(t)\,dt + \int_{t_1}^{t_2} G(t)\,dw_2(t), \qquad (5.19)$$

$(0 \le t_1 \le t_2 \le T)$ implies that $z(\cdot)$ is continuous, and that

$$\mathscr{E}\{z(t_2)\,|\,\mathcal{Y}_{t_1}\} = z(t_1) + \mathscr{E}\left\{\int_{t_1}^{t_2} F(t)\mathscr{E}\{e(t)\,|\,\mathcal{Y}_t\}\,dt\,\Big|\,\mathcal{Y}_{t_1}\right\}$$

$$= z(t_1);$$

that is, the $z(\cdot)$ process is a continuous martingale relative to the $\mathcal{Y}_t$. We show next that

$$\mathscr{E}\{[z(t_2) - z(t_1)][z(t_2) - z(t_1)]'\,|\,\mathcal{Y}_{t_1}\} = \int_{t_1}^{t_2} G(t)G(t)'\,dt \quad (5.20)$$

$(0 \le t_1 \le t_2 \le T)$. Note first, from (5.8) and (5.17), that

$$de = (A - QF'(GG')^{-1}F)e\,dt + C\,dw_1 - QF'(GG')^{-1}G\,dw_2. \quad (5.21)$$

For the process $\{z(t), e(t);\ t \ge t_1\}$ we introduce the matrix

$$R(t) = \begin{bmatrix} R_1(t) & R_3(t) \\ R_3(t)' & R_2(t) \end{bmatrix}$$

$$= \mathscr{E}\left\{ \begin{bmatrix} z(t)z(t)' & z(t)e(t)' \\ e(t)z(t)' & e(t)e(t)' \end{bmatrix} \,\Big|\, \mathcal{Y}_{t_1} \right\}$$

From the linear equations (5.18), (5.21), and the fact that the $dw_i(t)$ increments are independent of $\mathcal{Y}_{t_1}$ if $t \ge t_1$, it is immediately verified that $R(t)$ is given by the linear differential equations

$$dR_1/dt = FR_3 + R_3'F' + GG',$$

$$dR_2/dt = [A - QF'(GG')^{-1}F]R_2 + R_2[A - QF'(GG')^{-1}F]'$$
$$+ CC' + QF'(GG')^{-1}FQ,$$

$$dR_3/dt = [A - QF'(GG')^{-1}F]R_3 + (R_2 - Q)F',$$

$(t_1 \le t \le T)$, together with the initial condition

$$R(t_1) = \begin{bmatrix} z(t_1 z(t_1)' & 0 \\ 0 & Q(t_1) \end{bmatrix}.$$

By use of (5.15), $R_2 - Q$ satisfies the linear equation

$$d(R_2 - Q)/dt = [A - QF'(GG')^{-1}F](R_2 - Q)$$
$$+ (R_2 - Q)[A - QF'(GG')^{-1}F]'$$

$(t \geq t_1)$; and since $R_2(t_1) = Q(t_1)$ there follows $R_2(t) = Q(t)$, $t_1 \leq t \leq T$. This result implies in turn that $R_3(t) = 0$, $t_1 \leq t \leq T$, and therefore

$$R_1(t) = z(t_1)z(t_1)' + \int_{t_1}^{t} G(s)G(s)' \, ds,$$

$(t_1 \leq t \leq T)$. Finally, since

$$\mathcal{E}\{z(t_1)[z(t_2) - z(t_1)]' \mid \mathcal{Y}_{t_1}\} = z(t_1)\mathcal{E}\{[z(t_2) - z(t_1)]' \mid \mathcal{Y}_{t_1}\},$$
$$= 0,$$

the verification of (5.20) is complete. We now use the fact that a continuous martingale whose covariance satisfies a relation of type (5.20) is actually a (suitably scaled) Wiener process (cf. Doob [10, p. 384, Theorem 11.9]). That is, there exists a standard Wiener process $\{\hat{w}(t): 0 \leq t \leq T\}$, carried by the basic probability space $(\Omega, \mathcal{F}, \mathbf{P})$, and such that

$$dz(t) = [G(t)G(t)']^{1/2} \, d\hat{w}(t), \qquad t \in [0, T] \tag{5.22}$$

Combining (5.17), (5.18), and (5.22) we get finally,

$$d\hat{x} = A\hat{x} \, dt + \beta(t) \, dt + QF'(GG')^{-1/2} \, d\hat{w}(t),$$
$$\hat{x}(0) = \mathcal{E}\{x_0\}.$$

Suppose now that

$$u(t) = \hat{\psi}[t, \hat{x}(t)],$$

where $\hat{\psi} \in \hat{\Psi}$. Under the regularity conditions (5.6) and A.4, the Itô equation

$$d\hat{x} = A\hat{x} \, dt + b[t, \hat{\psi}(t, \hat{x})] \, dt + QF'(GG')^{-1/2} \, d\hat{w}$$
$$\hat{x}(0) = \mathcal{E}\{x_0\} \tag{5.23}$$

determines a diffusion process on $[0, T]$. Let $\xi \in R^n$ denote a value of $\hat{x}$ and let $V: R^n \to R^1$ have continuous derivatives up to second order. The differential generator of the $\hat{x}$ process is the elliptic operator $\mathcal{L}(\hat{\psi})$ given by (2.11):

$$\mathcal{L}(\hat{\psi})V(\xi) \equiv \tfrac{1}{2} \operatorname{tr}\{\hat{C}'V_{\xi\xi}(\xi)\hat{C}\} + (A\xi + b[t, \hat{\psi}(t, \xi)])'V_\xi(\xi). \tag{5.24}$$

In (5.24) $\hat{C} = QF'(GG')^{-1/2}$ and $V_\xi(V_{\xi\xi})$ denotes the vector (matrix) of first (second) partial derivatives of V.

Next, we show that $\mathscr{L}$ is uniformly elliptic. Observe first that $Q(t) > 0$; in fact (5.15) can be written

$$dQ/dt = \hat{A}Q + Q\hat{A}' + CC', \tag{5.25}$$

where

$$\hat{A} = A - \tfrac{1}{2}QF'(GG')^{-1}F.$$

If $\mathscr{A}(t)$ is the matrix determined by

$$d\mathscr{A}(t)/dt = \hat{A}(t)\mathscr{A}(t), \qquad \mathscr{A}(0) = I,$$

then (4.25) and A.7 imply

$$Q(t) \geq \mathscr{A}(t)Q_0\mathscr{A}(t)'$$
$$\geq c_{10}I, \qquad t \in [0, T].$$

This fact, combined with A.2 and A.3 shows that

$$\hat{C}(t)\hat{C}(t)' \geq c_{11}I, \qquad t \in [0, T]. \tag{5.26}$$

We verify now that controls of the class $\hat{\mathscr{U}}$, i.e.,

$$u(t) = \hat{\psi}[t, \hat{x}(t)], \qquad \hat{\psi} \in \hat{\Psi}, \tag{5.27}$$

are admissible. In view of (5.6) it is enough to check that $\hat{x}(t)$ is a uniformly Lipschitz-continuous functional of $\pi_t y$. By A.1 and (5.15) the matrix

$$K(t) = Q(t)F(t)'[G(t)G(t)']^{-1},$$

is continuously differentiable in $[0, T]$, so that

$$\int_0^t K(s) \, dy(s) = K(t)y(t) - \int_0^t (dK(s)/ds)y(s) \, ds. \tag{5.28}$$

Clearly the right side of (5.28) is a continuous functional of $\pi_t y$. Consider now (5.17) with $\beta(t) = b[t, \hat{\psi}(t, \hat{x}(t))]$. Integrating (5.17) and substituting (5.28), one obtains for $\hat{x}(\cdot)$ a Volterra equation with kernel uniformly Lipschitzian in the variable $\hat{x}$. From this it is easy to establish the required Lipschitz continuity in $\mathscr{C}$ of the mapping $\pi_t y(\cdot) \to \hat{x}(\cdot)$.

To conclude this subsection, we verify that the conditional distribution of $x(t)$ given $\mathscr{Y}_t$ is Gaussian and that, if $0 \leq t_1 \leq t_2 \leq t_3 \leq T$, the increments $\hat{w}(t_3) - \hat{w}(t_2)$ are independent of $\mathscr{Y}_{t_1}$. These facts are crucial to the later development. To prove the first statement recall that

$x(t) = \tilde{x}(t) + x^*(t)$ where $x^*(t)$ is $\mathscr{Y}_t$-measurable. Furthermore, by the linearity of (5.9) and (5.12b), the conditional distribution of $\tilde{x}(t)$ given $\tilde{\mathscr{Y}}_t$ is Gaussian. Since $\tilde{\mathscr{Y}}_t = \mathscr{Y}_t$ the assertion follows. Next, the relation $e(t) = x(t) - \hat{x}(t)$ shows that the conditional distribution of $e(t)$ given $\mathscr{Y}_t$ is Gaussian, with covariance $Q(t)$ depending only on t. Also, combining (5.18) and (5.21) we see that $z(t_2) - z(t_1)$ can be written as a linear functional of $e(t_1)$ and the $w_i(s)$ increments ($i = 1, 2$) for $t_1 \le s \le t_2$. This implies that the conditional distribution of $z(t_3) - z(t_2)$ given $\mathscr{Y}_{t_1}$ is Gaussian, with zero mean, and covariance independent of $\mathscr{Y}_{t_1}$; hence the same is true of $\hat{w}(t_3) - \hat{w}(t_2)$, and the independence assertion follows. Finally, if the Itô equation (5.23) holds then $\hat{x}(t_2)$ is measurable relative to the sample space of $\hat{x}(t_1)$ and the $\hat{w}(s)$ increments for $t_1 \le s \le t_2$. Thus if $f: R^n \to R^1$ is an arbitrary bounded measurable function, then

$$\mathscr{E}\{f[\hat{x}(t_2)] \mid \mathscr{Y}_{t_1}\} = \mathscr{E}\{f[\hat{x}(t_2)] \mid \hat{x}(t_1)\}, \tag{5.29}$$

with probability 1.

E. A Sufficient Condition for Optimality

Let $\mathscr{G}(x; t, \xi)$ be the Gaussian probablity density in R^n with mean ξ and covariance matrix $Q(t)$:

$$\mathscr{G} = (2\pi)^{-n/2} [\det Q(t)]^{-1/2} \exp[-\tfrac{1}{2}(x - \xi)' Q(t)^{-1} (x - \xi)]. \tag{5.30}$$

By the results of Section V.D, if u is a fixed vector of U

$$\hat{L}(t, \xi, u) \equiv \mathscr{E}\{L[t, x(t), u] \mid \hat{x}(t) = \xi\}$$
$$= \int_{R^n} L(t, x, u) \mathscr{G}(x; t, \xi) \, dx.$$

It is verified in Section V.F that $\hat{L}$ satisfies the conditions imposed on L in A.5. On this assumption we establish the following optimality criterion (cf. Krasovskii [42] and Wonham [74]).

Lemma 5.2 (Optimality Criterion). *Suppose there exist an element* $\hat{\psi}^0 \in \hat{\Psi}$ *and a function* $V: [0, T] \times R^n \to R^1$ *such that*

(i) $V, V_t, V_\xi, V_{\xi\xi}$ *are continuous, and*

$$|V| + |V_t| + |\xi| \, |V_\xi| + |V_{\xi\xi}| \le c_{12}(1 + |\xi|^2). \tag{5.31}$$

(ii)

$$0 = V_t(t, \xi) + \hat{\mathcal{L}}(\hat{\psi}^0)V(t, \xi) + \hat{L}[t, \xi, \hat{\psi}^0(t, \xi)] \qquad (5.32a)$$

$$0 \le V_t(t, \xi) + \hat{\mathcal{L}}(u)V(t, \xi) + \hat{L}(t, \xi, u) \qquad (5.32b)$$

for all $(t, \xi, u) \in [0, T] \times R^n \times U.$

$$V(T, \xi) = 0, \qquad \xi \in R^n. \qquad (5.32c)$$

Then the control $u = \hat{\psi}^0$ *is optimal in* $\mathcal{U}$.

For the proof, introduce the random variable

$$W(t) = \mathcal{E}\left\{ \int_t^T L[s, x(s), \hat{\psi}^0(s, \hat{x}(s))\, ds] \,|\, \mathcal{Y}_t \right\}$$

where $x(t)$ is the solution of (5.1) with $u(t) = \hat{\psi}^0[t, \hat{x}(t)]$, and $\hat{x}(t)$ is given by (5.23) with $\hat{\psi} = \hat{\psi}^0$. Now

$$W(t) = \mathcal{E}\left\{ \int_t^T \mathcal{E}\{L[s, x(s), \hat{\psi}^0(s, \hat{x}(s))] \,|\, \mathcal{Y}_s\}\, ds \,|\, \mathcal{Y}_t \right\}$$

$$= \mathcal{E}\left\{ \int_t^T \hat{L}[s, \hat{x}(s), \hat{\psi}^0(s, \hat{x}(s))]\, ds \,|\, \mathcal{Y}_t \right\}$$

$$= \mathcal{E}\left\{ \int_t^T \hat{L}[s, \hat{x}(s), \hat{\psi}^0(s, \hat{x}(s))]\, ds \,|\, \hat{x}(t) \right\}, \qquad (5.33)$$

where we have used (5.29). By (5.31), (5.32a), (5.33), and Itô's integration formula (2.15)

$$W(t) = -\mathcal{E}\left\{ \int_t^T (V_t[s, \hat{x}(s)] + \hat{\mathcal{L}}(\hat{\psi}^0)V[s, \hat{x}(s)])\, ds \,|\, \hat{x}(t) \right\}$$

$$= V[t, \hat{x}(t)].$$

In particular,

$$V[0, \hat{x}(0)] = J[\hat{\psi}^0]. \qquad (5.34)$$

To show that $\hat{\psi}^0$ is optimal, let $u(t)$ be an arbitrary control

$$u(t) = \psi(t, \pi_t y),$$

where $\psi \in \Psi$; *and now write* $x(t)$, $\hat{x}(t)$ *for the corresponding solution of* (5.1) and (5.17). Since the second moment $\mathcal{E}\{|x(t)|^2\}$ is integrable on $[0, T]$ (see Fleming and Nisio [18]) and since (5.31) holds, we may apply Itô's generalized integration formula (2.15)*, to obtain

$$\mathscr{E}\{V[t, \hat{x}(t)] \,|\, \mathscr{Y}_t\}$$

$$= -\mathscr{E}\left\{\int_t^T [V_s[s, \hat{x}(s)] + \hat{\mathscr{L}}(\psi)V[s, \hat{x}(s)]] \, ds \,|\, \mathscr{Y}_t\right\}$$

$$\le -\mathscr{E}\left\{\int_t^T (V_s[s, \hat{x}(s)] + \hat{\mathscr{L}}(\hat{\psi}^0)V[s, \hat{x}(s)]\right.$$

$$\left. + \hat{L}[s, \hat{x}(s), \hat{\psi}^0(s, \hat{x}(s))] - \hat{L}[s, \hat{x}(s), \psi(s, \pi_s y)]) \, ds \,|\, \mathscr{Y}_t\right\}$$

$$= \mathscr{E}\left\{\int_t^T \hat{L}[s, \hat{x}(s), \psi(s, \pi_s y)] \, ds \,|\, \mathscr{Y}_t\right\}$$

$$= \mathscr{E}\left\{\int_t^T L[s, x(s), \psi(s, \pi_s y)] \, ds \,|\, \mathscr{Y}_t\right\} \tag{5.35}$$

Here the inequality results from (5.32b) with $u = \psi$ in the right side, and the last equality follows as in (5.33). Setting $t = 0$ in (5.35), and using (5.3) and (5.34), we get

$$J[\hat{\psi}^0] \le J[\psi].$$

This inequality states that $\hat{\psi}^0$ is optimal.

It remains to prove that $\hat{\psi}^0$ and V exist. We shall do this by solving Bellman's equation.

F. SOLUTION OF BELLMAN'S EQUATION

Observe that (5.32) is formally equivalent to Bellman's functional equation

$$\min_{u \in U}[V_t(t, \xi) + \hat{\mathscr{L}}(u)V(t, \xi) + \hat{L}(t, \xi, u)] = 0, \qquad (t, \xi) \in [0, T] \times R^n, \tag{5.36a}$$

$$V(T, \xi) = 0. \tag{5.36b}$$

The minimization in (5.36a) is to be done at each fixed $(t, \xi) \in [0, T] \times R^n$. It will be shown first that this is possible. Write $V_\xi = p$ and consider the function

$$\lambda(t, \xi, p, u) = b(t, u)'p + \hat{L}(t, \xi, u), \tag{5.37}$$

defined for $(t, \xi, p, u) \in [0, T] \times R^n \times \{p : |p| \le \pi\} \times U$.

We shall verify that $\hat{L}$ satisfies the conditions imposed on L in A.5.

Clearly $\hat{L}$ is bounded. From the elementary relations

$$|e^a - e^b| \le \tfrac{1}{2}|a - b|(e^a + e^b),$$

$$|a^{-1/2} - b^{-1/2}| = (ab)^{-1/2}(a^{1/2} + b^{1/2})^{-1}|a - b|, \quad (a, b > 0),$$

$$|ac - bd| \le |c||a - b| + |b||c - d|;$$

the fact that

$$c_{12}I \le Q(t) \le c_{13}I, \quad t \in [0, T];$$

and the continuity of dQ/dt; there results

$$|\mathscr{G}(x; t_1, \xi_1) - \mathscr{G}(x; t_2, \xi_2)|$$
$$\le c_{14}(|\xi_1 - \xi_2| + |t_1 - t_2|)[\exp\{-c_{15}|x - \xi_1|^2\}$$
$$+ \exp\{-c_{15}|x - \xi_2|^2\}].$$

The assertion now follows by simple computations.[4] Henceforth $\hat{A}.5$ denotes A.5 with L, x replaced by $\hat{L}, \xi$.

Next, observe that the inequality in A.6 can be integrated over $x \in R^n$ with respect to $\mathscr{G}$, to yield

$\hat{A}.6 \quad [b(t, u)'p + \hat{L}(t, \xi, u)]_{uu} \ge c_{16}I.$

By virtue of $\hat{A}.5$, $\hat{A}.6$ the problem of minimizing with respect to u the function λ in (5.37) has the following solution.

Lemma 5.3. *There exists a (unique) function $\mu(t, \xi, p)$ with values in U such that*

(i) $\lambda[t, \xi, p, \mu(t, \xi, p)] \le \lambda(t, \xi, p, u)$ *for all*

$$(t, \xi, p, u) \in [0, T] \times R^n \times \{p: |p| \le \pi\} \times U.$$

(ii) μ *is uhc(α)in t, and is ulc in (ξ, p) in the domain*

$$(t, \xi, p) \in [0, T] \times R^n \times \{p: |p| \le \pi\}.$$

This result is due to Fleming [16, Lemma 2.1].
Write

$$\Lambda(t, \xi, p) = \xi'A(t)'p + \lambda[t, \xi, p, \mu(t, \xi, p)].$$

[4] These show that the Lipschitz condition in x on L, L_u could be relaxed.

With the substitution $u = \mu$ in (5.32) we obtain the semilinear parabolic equation

$$V_t(t, \xi) + \tfrac{1}{2} \operatorname{tr}\{\hat{C}(t)'V_{\xi\xi}(t, \xi)\hat{C}(t)\} + \Lambda[t, \xi, V_\xi(t, \xi)] = 0,$$

$$(t, \xi) \in [0, T] \times R^n$$

$$V(T, \xi) = 0, \qquad \xi \in R^n. \tag{5.38}$$

It remains to verify that the Cauchy problem (5.38) has a suitably smooth solution V. This conclusion follows by a theorem of Ladizhen-skaya et al. [47, p. 564, Theorem 8.1]. For ease of reference we check the hypotheses of the theorem in detail; page numbers refer to the book just cited.

(i) Condition (b), p. 564: by (5.37) and boundedness of μ, $\hat{L}$,

$$\Lambda(t, \xi, 0)V = \hat{L}[t, \xi, \mu(t, \xi, 0)]V \le c_{16} V^2 + c_{17}$$

for all $(t, \xi, V) \in [0, T] \times R^n \times R^1$.

(ii) Condition (b), p. 513: by A.1, A.4, $\hat{A}$.5, and Lemma 5.3, Λ is continuous in (t, ξ, p); as shown in V.D, $\hat{L}$ is uniformly elliptic; and it is clear that $\hat{C}\hat{C}'$ is bounded on $[0, T]$. Finally

$$|\Lambda(t, \xi, p| \le c_{18}(R)(1 + |p|)$$

in every cylinder $t \in [0, T]$, $|\xi| \le R$.

(iii) Condition (c), p. 513: by A.1, A.4, $\hat{A}$.5, and Lemma 5.3, $\Lambda(t, \xi, p)$ is uhc(α) in t, in every domain $t \in [0, T]$, $|\xi| \le R$, $|p| \le \pi$; and is ulc in ξ (resp. p) in the domain $t \in [0, T]$, $|p| \le \pi$ (resp. $t \in [0, T]$, $|\xi| \le R$). By A.1 and (5.15) it is clear also that $\hat{C}(t)\hat{C}(t)'$ is Hölder (even Lipschitz) continuous on $[0, T]$.

It follows by Theorem 8.1 of Ladizhenskaya et al. [47] that (5.38) has a solution $V(t, \xi)$, defined and bounded for $(t, \xi) \in [0, T] \times R^n$, such that V, V_t, V_ξ and $V_{\xi\xi}$ are uhc(α) in t and uhc(2α) in ξ, in every finite cylinder $t \in [0, T]$, $|\xi| \le R$. We will show that V_ξ is actually ulc in ξ on $[0, T] \times R^n$. It is enough to show that $V_{\xi\xi}$ is bounded. To this end, introduce the change of variables

$$\eta = S(t)\xi, \qquad t = t,$$

where S is uniquely determined by

$$dS(t)/dt = -S(t)A(t), \qquad t \in [0, T],$$

$$S(0) = I. \tag{5.39}$$

Setting $\tilde{V}(t, \eta) = V(t, S(t)^{-1}\eta)$, we obtain

$$\tilde{V}_t + \tfrac{1}{2} \operatorname{tr}\{\hat{C}(t)'S(t)\tilde{V}_{\eta\eta}S(t)\hat{C}(t)\} + \tilde{\Lambda}(t, \eta, \tilde{V}_\eta) = 0, \qquad t \in [0, T], \quad (5.40\text{a})$$

$$\tilde{V}(T, \eta) = 0, \qquad\qquad (5.40\text{b})$$

where

$$\tilde{\Lambda}(t, \eta, \tilde{p}) = \lambda[t, \xi, p, \mu(t, \xi, p)]\Big|_{\substack{\xi = S^{-1}\eta \\ p = S'\tilde{p}}}.$$

Observe first that $\tilde{\Lambda}(t, \eta, \tilde{p})$ is uhc(α) in t and ulc in $(\eta, \tilde{p})$, in its domain

$$(\eta, \tilde{p}) \in R^n \times S(t)'^{-1}\{p : |p| \le \pi\}, \qquad 0 \le t \le T.$$

The operator in (5.40) is uniformly parabolic on $[0, T] \times R^n$. Consider the fundamental solution corresponding to the operator defined by the first two terms in (5.40a). By direct calculation it follows, since $\tilde{\Lambda}$ is bounded, that $\tilde{V}_\eta$ is bounded, and π can be chosen *a priori* such that

$$|V_\xi(t, \xi)| \le \pi, \qquad (t, \xi) \in [0, T] \times R^n. \qquad (5.41)$$

Again, boundedness of $\tilde{\Lambda}$ implies that $\tilde{V}_\eta(t, \eta)$ is uhc in η on $[0, T] \times R^n$ (cf. Friedman [21, p. 193, Lemma 2]). Hence $\tilde{\Lambda}[t, \eta, \tilde{V}_\eta(t, \eta)]$ is uhc on $[0, T] \times R^n$. But then one sees easily (cf. Ladizhenskaya *et al.* [47, p. 308]) that $\tilde{V}_{\eta\eta}$ is uniformly bounded, hence so is $V_{\xi\xi}$.

With $V_\xi(t, \xi)$ ulc in ξ and Hölder continuous in t, we have that

$$\hat{\psi}^0(t, \xi) \equiv \mu[t, \xi, V_\xi(t, \xi)], \qquad (5.42)$$

enjoys the same properties, i.e., $\hat{\psi}^0 \in \hat{\Psi}$.

Returning to (5.40) and using as before the fundamental solution for the linear part, one obtains that $\tilde{V}_t$ is bounded, so that

$$|V_t| = |\tilde{V}_t - \xi'A'S'\tilde{V}_\eta| \le c_{19}(1 + |\xi|).$$

It is now clear that $V(t, \xi)$ and $\hat{\psi}^0(t, \xi)$ satisfy the conditions of Lemma (5.2). The proof of Theorem 5.1 is complete.

G. Linear Regulator

If Bellman's equation can be solved explicitly for functions V and $\hat{\psi}^0$ which satisfy the hypotheses of Lemma 5.1 then, of course, many of the restrictive conditions imposed in the general discussion become irrelevant. A well-known example is the following (cf. Joseph and Tou [32], and Wonham [74]). In (5.1) let

$$b[t, u(t)] = B(t)u(t);$$

let u range over R^m; and let

$$L(t, x, u) = x'M(t)x + u'N(t)u,$$

where $M(t)$ and $N(t)$ are, respectively, positive semidefinite and positive definite, with $N(t)^{-1}$ bounded on $[0, T]$.

We find then that

$$\hat{L}(t, \xi, u) = \xi'M(t)\xi + u'N(t)u + \text{tr}[M(t)Q(t)],$$

and Bellman's equation (5.36a) becomes

$$V_t + \tfrac{1}{2}\text{tr}(\hat{C}'V_{\xi\xi}\hat{C}) + \xi'A'V_\xi - \tfrac{1}{4}V_\xi'BN^{-1}B'V_\xi + \xi'M\xi + \text{tr}(MQ) = 0,$$
(5.43a)

$$V(T, \xi) = 0.$$
(5.43b)

Equation (5.43) has a solution of form

$$V(t, \xi) = p(t) + \xi'P(t)\xi.$$
(5.44)

Indeed substitution yields, for $0 \le t \le T$,

$$(dP/dt) + A'P + PA - PBN^{-1}B'P + M = 0,$$
$$P(T) = 0,$$
(5.45)

$$(dp/dt) + \text{tr}(\hat{C}'P\hat{C}) + \text{tr}(MQ) = 0,$$
$$p(T) = 0.$$
(5.46)

Equation (5.44) is the Riccati equation which appeared in connection with the linear regulator problem of Section III (cf. (3.11c)). The control law is given by

$$\begin{aligned}
u &= \phi(t, \hat{x}) \\
&= -\tfrac{1}{2}N(t)^{-1}B(t)'V_\xi(t, \hat{x}) \\
&= -N(t)^{-1}B(t)'P(t)\hat{x}.
\end{aligned}$$

Observe that the control law is the same function of $\hat{x}$ as it would be of x if channel noise were absent: We obtain complete separation of the problems of control and filtering.

A simple example will bring out the respective roles of dynamic noise $\dot{w}_1$ and channel noise $\dot{w}_2$. Let $n = m = 1$, $M = N = 1$, and consider the problem of minimizing

$$\mathscr{E}\left\{\int_0^T [x(t)^2 + u(t)^2]\, dt\right\},$$

for the system

$$dx = u\,dt + c\,dw_1$$
$$dy = x\,dt + g\,dw_2 \qquad (c, g \text{ are constants}).$$

Setting $\text{cov}\{x(t) \mid \mathscr{Y}_t\} = Q(t)$ we find from (5.15),

$$Q(t) = cg \cdot \frac{cg\,\sinh(ct/g) + q_0\,\cosh(ct/g)}{cg\,\cosh(ct/g) + q_0\,\sinh(ct/g)}.$$

From (5.44) there follows

$$P(t) = \tanh(T - t).$$

Then

$$p(0) = \int_0^T \{\hat{c}(s)^2 P(s) + Q(s)\}\,ds$$

$$= \int_0^T \{g^{-2} Q(s)^2 P(s) + Q(s)\}\,ds$$

$$\sim (c^2 + cg)T, \qquad T \to \infty.$$

VI. Linear Regulator with Randomly Jumping Parameters

A. Introduction

In this chapter we solve a linear regulator problem for a model of the type described in Section II.E. As in Section III it will be assumed that complete state information is available to the controller and that an optimal feedback control is to be sought. This problem was also studied by Krasovskii and Lidskii [45].

Let $\{y(t): t_0 \le t \le T\}$ be a continuous time Markov chain with differential transition matrix Q and state space $S = \{1, \ldots, v\}$. Consider the dynamic system

$$dx(t)/dt = A[t, y(t)]x(t) - B[t, y(t)]u(t), \qquad t_0 \le t \le T. \quad (6.1)$$

In (6.1), $x \in R^n$, $u \in R^m$ and for each $j \in S$, the matrices $A(\cdot, j)$, $B(\cdot, j)$ are continuous in $[t_0, T]$. Let Φ be the class of functions

$$\phi: [t_0, T] \times R^n \times S \to R^m,$$

such that, for some constant k (depending on ϕ),

$$|\phi(t, x, y) - \phi(t, \tilde{x}, y)| \le k|x - \tilde{x}|, \qquad \phi(t, x, y) \le k(1 + |x|),$$

for all t, x, $\tilde{x}$, y. We adopt Φ as the class of admissible controls. If (6.1) is supplemented by initial conditions

$$x(t_0) = x_0, \qquad y(t_0) = y_0,$$

where x_0 is independent of the $y(t)$ process, then the joint process $\{x(t), y(t): t_0 \leq t \leq T\}$, determined by setting

$$u(t) = \phi[t, x(t), y(t)],$$

is a Markov process (see Section II.E). It is not true in general that the $x(t)$ process alone is Markov.

B. LINEAR REGULATOR

Consider the problem of minimizing

$$J[u] = \mathscr{E}\left\{\int_{t_0}^{T} L[s, x(s), y(s), u(s)] \, ds\right\} \tag{6.2}$$

for a suitable function L. Using the results stated in Section II.E, and proceeding exactly as in Section III, we introduce

$V(t, x, y)$

$$= \min_{\phi \in \Phi} \mathscr{E}\left\{\int_{t}^{T} L[s, x(s), y(s), \phi\{s, x(s), y(s)\}] \, ds \,\middle|\, x(t) = x, y(t) = y\right\},$$

and are led to Bellman's equation

$$\min_{u \in R^m} \{V_t(t, x, y) + \mathscr{L}_u V(t, x, y) + L(t, x, y, u)\} = 0$$

$$(t, x, y) \in [t_0, T] \times R^n \times S; \quad (6.3a)$$

$$V(T, x, y) = 0, \qquad (x, y) \in R^n \times S. \tag{6.3b}$$

Here

$$\mathscr{L}_u V(t, x, i) \equiv [A(t, i)x - B(t, i)u]' V_x(t, x, i) - q_i V(t, x, i)$$

$$+ \sum_{\substack{j=1 \\ j \neq i}}^{v} q_{ij} V(t, x, j).$$

In particular, let

$$L(t, x, y, u) = x' M(t, y)x + u' N(t, y)u, \tag{6.4}$$

where $M(\cdot, y)$, $N(\cdot, y)$ are continuous on $[t_0, T]$, $M \geq 0$ and $N(t, y)$ $\geq \varepsilon I$, $t \in [t_0, T]$, $y \in S$, for some $\varepsilon > 0$. Set

$$V(t, x, y) = x'P(t, y)x, \qquad (6.5)$$

and write A_i for $A(\cdot, i)$, etc. Then we find that (6.3) holds provided

$$(dP_i/dt) + (A_i - B_i K_i)'P_i + P_i(A_i - B_i K_i) - q_i P_i + \sum_{\substack{j=1 \\ j \neq i}}^{v} q_{ij} P_j$$

$$+ M_i + K_i'N_i K_i = 0, \qquad i \in S, \quad t_0 \leq t \leq T, \qquad (6.6a)$$

$$K_i = N_i^{-1} B_i P_i, \qquad (6.6b)$$

$$P_i(T) = 0, \qquad i \in S. \qquad (6.6c)$$

It is not difficult to show that (6.6) has a unique solution $\{P_i(t),\ i = 1, \ldots, v\}$ on $[t_0, T]$: for (6.6a,b) comprise a set of coupled matrix Riccati equations which can be solved by quasi-linearization and successive approximation, just as in Wonham [81]. It may be helpful to summarize the technique. Suppose $\{P_i^0\}$ satisfies (6.6), let $\{\tilde{K}_i\}$ be an arbitrary set of v $m \times n$ matrices continuous on $[t_0, T]$, and let $\{\tilde{P}_i\}$ be the corresponding solution of the linear equation (6.6a,c) with K_i replaced by $\tilde{K}_i$. Set $K_i^0 = N^{-1} B_i P_i^0$. The identity

$$(A_i - B_i K_i^0)'P_i^0 + P_i^0(A_i - B_i K_i^0) + K_i^{0\prime} N_i K_i^0$$
$$= (A_i - B_i K_i)'P_i^0 + P_i^0(A_i - B_i K_i) + K_i'N_i K_i$$
$$\quad - (K_i - K_i^0)'N_i(K_i - K_i^0), \qquad (6.7)$$

shows that K_i^0 minimizes the left side of (6.6a), regarded as a function of K_i. Let $Q_i = \tilde{P}_i - P_i^0$. From (6.6) and (6.7) there follows

$$(dQ_i/dt) + (A_i - B_i \tilde{K}_i)'Q_i + Q_i(A_i - B_i \tilde{K}_i) - q_i Q_i + \sum_{j \neq i} q_{ij} Q_j \leq 0,$$

$$i \in S, \quad t \in [t_0, T].$$

$$Q_i(T) = 0, \qquad i \in S. \qquad (6.8)$$

Write $\Phi_i(t, s)$ for the fundamental matrix associated with $A_i - B_i \tilde{K}_i - \frac{1}{2} q_i I$. From (6.8) there results

$$Q_i(t) \geq \int_t^T \Phi_i(T, s)' \left[\sum_{j \neq i} q_{ij} Q_j(s) \right] \Phi_i(T, s)\, ds. \qquad (6.9)$$

The Volterra inequality (6.9) can be solved by successive approximations. Since $q_{ij} \geq 0$, the sequence $Q_i^{(r)}$, determined by setting $Q_i^{(0)} = 0$, is monotone nondecreasing, and this shows that $Q_i(t) \geq 0$, or

$$P_i^0(t) \leq \tilde{P}_i(t), \qquad i \in S, \quad t \in [t_0, T]. \qquad (6.10)$$

Inequality (6.10) expresses the *minimum* property of the solution of (6.6).

Now construct sequences $K_i^{(r)}$, $P_i^{(r)}(r = 1, 2, \ldots)$ as follows: $P_i^{(r)}$ is the solution of (6.6a,c) with $K_i = K_i^{(r)}$; $K_i^{(r+1)} = N_i^{-1}B_i'P_i^{(r)}$; $K_i^{(1)}$ is chosen arbitrarily. By the method just used one finds easily that

$$0 \le P_i^{(r+1)}(t) \le P_i^{(r)}(t), \qquad (r = 1, 2, \ldots),$$

and from this it follows that

$$P_i^0(t) \equiv \lim_{r \to \infty} P_i(t),$$

exists. Finally, an application of the dominated convergence theorem shows that $P_i^0(t)$ satisfies (6.6).

The optimal feedback control, obtained from (6.3a), is

$$\phi^0(t, x, i) = K_i^0(t)x = N_i(t)^{-1}B_i(t)'P_i^0(t)x. \tag{6.11}$$

Control is thus linear in the dynamic state x, but with "gain" parameters which depend on the current state of the step process $y(t)$. In applications the $y(t)$ process would represent parametric variations in linear dynamics, and such variations are usually not directly accessible to physical measurement. For this reason, the model described is currently of limited practical interest. Nevertheless, the result may provide a useful reference in future studies of adaptive systems, in which parametric variations are estimated from behavior of the dynamic state.

C. Asymptotic Behavior as $T - t_0 \to \infty$

It is of interest to determine under what conditions the solution $\{P_i(t)\}$ of (6.6) tends to a finite limit as $t \downarrow -\infty$. We shall see that existence of the limit reflects a stability property of the system (6.1). Throughout this section the matrices A_i, B_i, M_i, N_i are assumed to be independent of t. We set $T = 0$ and consider (6.6) for $-\infty < t \le 0$.

Intuition suggests that the behavior of $P_i(t)$ will be regular as $t \downarrow -\infty$ if the pairs (A_i, B_i) are stabilizable and if parameter jumps occur with sufficiently small mean frequency. In analogy to (3.18), one is led to the condition

$$\max_{i \in s} \inf_K \left| q_i \int_0^\infty \exp[-q_i\tau]\exp\{\tau(A_i - B_iK)'\}\exp\{\tau(A_i - B_iK)\} \, dt \right| < 1 \tag{6.12}$$

The main result is the following.

Theorem 6.1. *Suppose*

(i) *the pairs (A_i, B_i) are stabilizable $(i \in S)$*
(ii) *condition (6.12) holds*
(iii) *the pairs $(\sqrt{M_i}, A_i)$ are observable $(i \in S)$.*

Then the solution $\{P_i(t)\}$ of (2.5), defined for $t \leq 0$, has the properties

(i) $\bar{P}_i = \lim P_i(t)$ $(t \to -\infty)$ *exists $(i \in S)$*
(ii) $\{\bar{P}_i\}$ *is the unique solution of the quadratic system*

$$A_i'P_i + P_i A_i - P_i B_i' N^{-1} B_i P_i - q_i P_i + \sum_{\substack{j=1 \\ j \neq i}}^{v} q_{ij} P_j = 0, \qquad (i \in S),$$

$$(6.13)$$

such that $P_i \geq 0$, $i \in S$.

(iii) $\bar{P}_i > 0$ $(i \in S)$ *and the matrices*

$$\bar{C}_i \equiv A_i - B_i N_i^{-1} B_i' P_i - \tfrac{1}{2} q_i I,$$

are stable.

This result is a variation of Theorem 2.1 of Wonham [81] and its proof, along very similar lines, will only be outlined. We consider first the quadratic system (6.13), equivalent to (6.6a,b) with $dP_i/dt \equiv 0$. Choosing constant matrices K_i such that $A_i - B_i K_i$ is stable $(i \in S)$, we consider the linear equations

$$P_i = T_i(P_1, \ldots, P_v) + R_i, \qquad (i \in S), \qquad (6.14)$$

where

$$T_i(P_1, \ldots, P_v) = \int_0^\infty \exp[-q_i \tau] \exp[\tau(A_i - B_i K_i)'] \left(\sum_{\substack{j=1 \\ j \neq i}}^{v} q_{ij} P_j \right)$$

$$\times \exp[\tau(A_i - B_i K_i)] \, d\tau, \quad (6.15)$$

$$R_i = \int_0^\infty \exp[-q_i \tau] \exp[\tau(A_i - B_i K_i)']$$

$$\times (M_i + K_i' N_i K_i) \exp[\tau(A_i - B_i K_i)] \, d\tau. \quad (6.16)$$

Observe that (6.13) is true if (6.14) holds with K_i given by (6.6b) and provided the matrices

$$C_i \equiv A_i - B_i N_i^{-1} P_i - \tfrac{1}{2} q_i I$$

are stable. Now $T_i(P_1, \ldots, P_v) \geq 0$ $(i \in S)$ if $P_i \geq 0$ $(i \in S)$; also

$$T_i(P_1, \ldots, P_v) \leq q_i\left(\max_j |P_j|\right) \int_0^\infty \exp[\tau C_i'] \exp[\tau C_i] \, d\tau. \quad (6.17)$$

Choose, by (6.12), the K_i such that

$$T_i(P_1, \ldots, P_v) \leq \theta\left(\max_j |P_j|\right) I \quad (6.18)$$

for some $\theta \in (0, 1)$. With these K_i, the T_i are contractive, so (6.14) has a unique solution $P_i \geq 0$ $(i \in S)$. It is even true that $P_i > 0$, because observability of $(\sqrt{M_i}, A_i)$ implies observability of

$$((M_i + K_i'N_i K_i)^{1/2}, A_i - B_i K_i)$$

(Wonham [81, Lemma 2.2]), which in turn implies $R_i > 0$.

To solve (6.13) one constructs sequences $P_i^{(r)}$, $K_i^{(r)}$ according to (in obvious notation)

$$P_i^{(r)} = T_i(P_1^{(r)}, \ldots, P_v^{(r)}; K_i^{(r)}) + R_i(K_i^{(r)}),$$

$$K_i^{(r+1)} = N_i^{-1} B_i P_i^{(r)},$$

$(r = 1, 2, \ldots)$, with $K_i^{(1)}$ chosen so that (6.18) holds. Brief computations (cf. Wonham [81]) show that

$$0 \leq P_i^{(r+1)} \leq P_i^{(r)}, \quad (r = 1, 2, \ldots),$$

and therefore $P_i^* \equiv \lim P_i^{(r)}$ $(r \to \infty)$ exists $(i \in S)$. With $K_i^* = N_i^{-1} B_i P_i^*$ there follows

$$(A_i - B_i K_i^*)' P_i^* + P_i^*(A_i - B_i K_i^*) - q_i P_i^*$$

$$+ \sum_{\substack{j=1 \\ j \neq i}}^{v} q_{ij} P_j^* + M_i + K_i^{*'} N_i K_i^* = 0, \quad i \in S.$$

This equality, with assumption (iii), implies that the matrices

$$A_i - B_i K_i^* - \tfrac{1}{2} q_i I, \quad i \in S,$$

are stable (cf. Wonham [81], Lemma 4.2). Then, as above, (6.14) shows that $P_i^* > 0$. Uniqueness of P_i^* in the class $P_i \geq 0$ follows by an analog of the minimum property derived in Section VI.B.

Next we consider the Riccati equation (6.6) and show that the limits $\bar{P}_i$ exist. First, it is not difficult to show that

$$P_i(t) \geq P_i(t'), \quad i \in S, \quad (6.19)$$

if $t \leq t' \leq 0$. Let $P_i^*(t)$ $(i \in S)$ be the solution of (6.6a,c) with $K_i = K_i^*$. Using an integral representation of the linear equations (6.6a,c) one can verify that

$$P_i^*(t) \uparrow P_i^*, \qquad t \to -\infty. \tag{6.20}$$

By the minimum property of the solution of (6.6),

$$P_i(t) \leq P_i^*(t). \tag{6.21}$$

From (6.19)–(6.21), $P_i(t)$ is bounded and monotone on $(-\infty, 0]$, and this implies that $\bar{P}_i$ exists. Since (6.6) shows that the derivatives dP_i/dt, d^2P_i/dt^2 are bounded and continuous, convergence of the integral

$$\int_{-\infty}^{0} [dP_i(t)/dt] \, dt,$$

implies that $dP_i/dt \to 0$ $(t \to -\infty)$. The fact that $\bar{P}_i = P_i^*$ follows by passage to the limit $(t \to -\infty)$ in (6.6). The proof of Theorem (6.12) is complete.

Write $\bar{K}_i = N_i^{-1} B_i' \bar{P}_i$. It is possible to interpret the time-invariant feedback law

$$\bar{\phi}(x, y) = \bar{K}_i x, \qquad (y = i \in S), \tag{6.22}$$

as a minimizing control for the functional

$$\mathscr{E}\left\{ \int_0^\infty L[x(t), y(t), u(t)] \, dt \right\}. \tag{6.23}$$

Thus, consider (6.1), defined for $t \geq 0$. Write

$$\bar{\phi}(x, y) = L[x, y, \bar{\phi}(x, y)]$$

and x_t for $x(t)$, etc. By monotone convergence

$$\mathscr{E}_{x,y}\left\{ \int_0^\infty \bar{L}(x_s, y_s) \, ds \right\}$$

$$= \lim_{t \uparrow \infty} \mathscr{E}_{x,y}\left\{ \int_0^t \bar{L}(x_s, y_s) \, ds \right\}$$

$$= \lim x' P^*(-t, y) x \qquad (t \uparrow \infty)$$

$$= x' \bar{P}_i x, \qquad (y = i \in S) \tag{6.24}$$

Let $\phi(t, x, y)$ be an arbitrary control which satisfies the general conditions of Section VI.A, and let

$$\phi^0(s, x, y; t) = K_i^0(s - t)x, \qquad (y = i \in S),$$

where

$$K_i^0(\tau) = N_i^{-1} B_i' P_i(\tau), \qquad \tau \le 0.$$

Write

$$L(x_s, y_s) = L[x_s, y_s, \phi(s, x_s, y_s)]$$
$$L^0(x_s, y_s; t) = L[x_s, y_s, \phi^0(s, x_s, y_s; t)]$$

where in each case the x_s process is determined by the corresponding choice of ϕ. By optimality of ϕ^0 in the interval $0 \le s \le t$,

$$x'P(-t, y)x = \mathscr{E}_{x,y}\left\{\int_0^t L^0(x_s, y_s; t)\, ds\right\} \le \mathscr{E}_{x,y}\left\{\int_0^t L(x_s, y_s)\, ds\right\}$$

so that

$$x'\overline{P}(y)x \le \lim_{t\to\infty} \mathscr{E}_{x,y}\left\{\int_0^t L(x_s, y_s)\, ds\right\} \qquad (6.25)$$

Then (6.24) and (6.25) show that $\overline{\phi}$ is optimal in the sense that

$$\mathscr{E}_{x,y}\left\{\int_0^\infty \overline{L}(x_s, y_s)\, ds\right\} \le \lim_{t\to\infty} \mathscr{E}_{x,y}\left\{\int_0^t \overline{L}(x_s, y_s)\, ds\right\},$$

for arbitrary admissible ϕ.

It is interesting that, for the limiting optimal system, the matrices $A_i - B_i \overline{K}_i$ need not all be stable. Roughly speaking, if q_i is large, parameter states with $y(t) = i$ are of brief duration, and short "intervals of instability" can be tolerated. As an example, let $S = \{1, 2\}$, $n = m = 1$, and

$$dx/dt = [2y(t) - 3]x(t) - u(t),$$

$$\mathscr{E}\left\{\int_0^\infty [x(t)^2 + u(t)^2]\, dt\right\} = \min.,$$

$$Q = \begin{bmatrix} -q_1 & q_1 \\ q_2 & -q_2 \end{bmatrix}$$

The condition (6.12) is satisfied, and from (6.13)

$$q_1 = P_1(2 + P_1)/(P_2 - P_1),$$
$$q_2 = P_2(2 - P_2)/(P_2 - P_1).$$

The controls are

$$\phi(x, i) = \overline{K}_i x = P_i x, \qquad (i = 1, 2).$$

Thus $0 < P_1 < P_2 < 2$ for all $q_1, q_2 > 0$. Choosing $q_1 = 5$, $q_2 = 15/4$, we find

$$P_1 = 1/2, \qquad P_2 = 3/4.$$

The system matrices (of dimension 1×1) are

$$A_1 - B_1 K_1 = -1 - 1/2 = -3/2,$$
$$A_2 - B_2 K_2 = 1 - 3/4 = 1/4.$$

Thus the linear closed-loop system actually operates in an unstable mode for a fraction of time $q_1/(q_1 + q_2) = 4/7$. Of course, it is true that

$$\int_0^\infty \{x(t)^2 + \phi[x(t), y(t)]^2\}\, dt < \infty$$

with probability 1.

Finally, we note that the stability criterion (6.12) need not be satisfied, even though (A_i, B_i) is stabilizable. A numerical counterexample is

$$A_i = \begin{bmatrix} 2 & 1 \\ -7 & -3 \end{bmatrix}, \qquad B_i = 0, \qquad q_i = 4.$$

Here A_i is stable and the left side of (6.12) is $\gtrsim 4$. On the other hand, explicit necessary conditions for stability are not presently available and it is not known if (6.12) is especially stringent.

D. GENERALIZATIONS

The results of this chapter could be generalized in several directions. To the right side of (6.1) could be added terms representing Gaussian white noise, as in Section III. It should be clear how the solution of the generalized regulator problem can be carried through. As in Section III, it is natural to consider optimal stationary control. For this one is led to a stability criterion which incorporates conditions (3.18) and (6.12).

Finally, it is possible to extend the theory to the case where the $y(t)$ process has a continuous state space. However, this generalization leads to equations which are significantly more involved from the viewpoint of numerical computation.

VII. Some Current Problems

In this section we review briefly some problems of current interest, referring the reader to the literature for details.

A. Existence of Optimal Feedback Controls

A standard approach to the problem of existence of optimal feedback controls is to adopt the formulation of dynamic programming and investigate the existence of a solution of Bellman's functional equation. This method has been used repeatedly in preceding sections. It has the conceptual advantage of concreteness, yielding an explicit description of an optimal control, and is effective insofar as one can exploit the general existence theory of equations of the relevant type. In addition, dynamic programming may suggest an algorithm for computing an optimal control numerically, although insuperable difficulties of computation size are met in naive application of dynamic programming to most problems of realistic complexity.

For optimal processes of diffusion type we have seen that the question reduces to the study of various quasilinear parabolic or elliptic partial differential equations. General existence theorems have been given by Friedman [21] and Ladizhenskaya et al. [47]. Friedman's results were applied by Girsanov [23] to a minimax problem of two competing players in control of a diffusion; and by Fleming [14] to the problem of optimal passage to the boundary of a cylinder. An application of a theorem of Ladizhenskaya on the Cauchy problem was given in Section V; of course, the reasoning used there in proving the separation theorem also yields a result for the simpler case where full state information is available to the controller. Other applications of differential equation methods to the existence problem include Mandl's [50] study of optimal stationary control on a bounded (space) interval and [51] of optimal switched control of mean passage time for a Wiener process.

The foregoing results are classical in nature, requiring rather stringent hypotheses of smoothness on the coefficients of the differential equations involved. From an applications viewpoint perhaps the most serious assumption is that the elliptic or parabolic operators that occur be uniform: in physical terms that an independent white-noise perturbation act on each component of the dynamic state. By going to generalized solutions Fleming [16] discussed the degenerate parabolic case for a stochastic game, and with Nisio [18] for a Markovian control problem with one controller. However, an unresolved difficulty is to show that the (generalized) solution of Bellman's equation determines an optimal control for which the solution of the corresponding stochastic differential equation is defined.

An abstract existence theorem independent of results from partial differential equations has been given by Fleming and Nisio [18]. Here the method is probabilistic, making use of Prokhorov's [58] results on compact sets of measures on a metric space, and Skorokhod's [62] representation theorem for associated sets of random processes.

Finally, we mention a quite different approach, again due to Fleming [17], via the derivation of necessary conditions for existence of an optimal control in terms of stochastic Lagrange multipliers. These conditions lead to boundary problems for the forward and backward operators of the diffusion process, together with a maximum principle of the type of Pontryagin *et al.* [56]. Although of theoretical interest, such results have not yet led to alternative methods of determining an optimal control explicitly.

B. MISCELLANEOUS FEEDBACK CONTROL PROBLEMS

In this subsection we mention briefly some further classes of control problems which lead to random differential equations and which have been considered in the literature.

a. Extremal control systems. An extremal system is one whose performance is optimized by automatic search in a coordinate space of adjustable parameters. Stochastic analysis is called for when observations of the output or the input–output relation itself are subject to random disturbances. Then the quantities of interest include stationary variance of tracking error and mean passage time to the optimal parameter settings. By such criteria it is possible to optimize, in turn, the parameters (e.g., step size and rate) of the search.

The systematic study of stochastic extremal systems was begun by Feldbaum [13]. If the parameter step size of the search is "small," the search can be approximated by a diffusion process, for which the limiting equilibrium distribution has been discussed (Jacobs and Wonham [29]). A detailed account of the theory is given by Pervozvanskii [53].

b. Containment probabilities. Let $\{x(t): 0 \leq t \leq T\}$ be a separable stochastic process in R^n and let $\mathscr{D} \subset R^n$ be an open domain with smooth boundary $\partial \mathscr{D}$. Define

$$P(t, x) = P[x(s) \in \mathscr{D}, 0 \leq s \leq t \mid x(0) = x],$$

for $x \in \mathscr{D}$, $0 \leq t \leq T$. $P(t, x)$ is the *containment probability* of $x(\cdot)$ relative to $\mathscr{D}$ (cf. Friedland *et al.* [19]). In applications $x(t)$ is the state vector of a system subject to random disturbances and $\mathscr{D}$ is a tolerance domain within which the system state is supposed to be maintained by control. Typically a performance specification for the system might be $P(T, x_0) \geq 1 - \varepsilon$, where x_0 is a fixed state in $\mathscr{D}$ and $\varepsilon > 0$ is small.

As a special case suppose X_ϕ is the diffusion process determined by

$$dx = f(x, u)\, dt + G(x)\, dw, \qquad 0 \leq t \leq T,$$

$$u = \phi(t, x), \qquad \phi \in \Phi,$$

$$x(0) = x_0.$$

Here Φ is a suitable class of feedback controls, and we seek $\phi^0 \in \Phi$ such that

$$P(t, x; \phi^0) \geq P(t, x; \phi); \qquad \phi \in \Phi, \quad t \in [0, T], \quad x \in \mathscr{D}. \tag{7.1}$$

Using the definition of P together with the Markov property, we are led to Bellman's equation

$$\frac{\partial P(t, x)}{\partial t} + \max_u \mathscr{L}_u P(t, x) = 0, \qquad 0 \leq t \leq T, \quad x \in \mathscr{D}$$

$$P(0, x) = 1, \qquad x \in \mathscr{D}$$

$$P(t, x) = 0, \qquad 0 \leq t \leq T, \quad x \in \partial\mathscr{D} \tag{7.2}$$

To show formally that (7.2) is a sufficient condition for optimality observe that, if $\phi \in \Phi$, $P(t, x; \phi)$ is determined by the equation (cf. Dynkin [11; Chapter 13, §5])

$$\partial P/\partial t + \mathscr{L}_\phi P = 0, \qquad 0 \leq t \leq T, \quad x \in \mathscr{D}$$

with the initial-boundary conditions of (7.2); and that, if P^0 is the solution of (7.2),

$$0 = \frac{\partial P^0}{\partial t} + \mathscr{L}_{\phi^0} P^0 \geq \frac{\partial P^0}{\partial t} + \mathscr{L}_\phi P^0, \qquad 0 \leq t \leq T, \quad x \in \mathscr{D}, \quad \phi \in \Phi.$$

The inequality (7.1) follows immediately by application of the maximum principle for parabolic equations.

Really effective methods of attacking problems of the type of (7.2) are apparently not known; for interesting preliminary results in this direction see, however, Friedland *et al.* [19], and Van Mellaert [67]. The dual problem of entering a target domain from the exterior region has been explored by Luh [49].

c. Control of stochastic service systems. A relatively unexplored area of control theory is the optimal control of stochastic service systems, i.e., of systems in which either the demands for service or the services given, or both, have a stochastic nature (Riordan [59]). The problem can be regarded as that of optimizing "queue discipline," that is, the logic by which resources (services) are allocated to demand (customers, calls, etc. waiting to be served). Situations of this type are numerous: telephone, highway, and air traffic control; control of computation and information flow in time-shared computer systems; allocation of donated organs among potential recipients of organ transplants.

One general model which has received detailed study is the following, proposed by Howard [24, 25]. The state space $\mathscr{X}$ of the system is $\{1, 2, \ldots, N\}$ or $\{1, 2, \ldots\}$; the set of admissible controls α is $\mathscr{A}$. We make the special assumption that $\mathscr{A} = \mathscr{A}_1 \times \mathscr{A}_2 \times \cdots$ and write $\alpha = (\alpha_1, \alpha_2, \ldots)$, $\alpha_i \in \mathscr{A}_i$. For each fixed $\alpha \in \mathscr{A}$, the state $x(t)$ evolves as a right-continuous Markov step process (cf. Section II.E) with differential transition matrix $Q(\alpha)$, defined as in (2.24), where $q_{ij}(\alpha) = q_{ij}(\alpha_i)$. Thus α_i is the control acting on the $x(\cdot)$ process, for t such that $x(t) = i \in \mathscr{X}$. Associated with the state i and control α is a numerical cost $L_i(\alpha) = L_i(\alpha_i)$.

In the applications interest has centered on minimizing the stationary expected cost

$$\mathscr{E}_\alpha\{L[x(s), \alpha(x(s))]\},$$

with respect to admissible controls $\alpha \in \mathscr{A}$. Here $L[x, \alpha(x)] = L_i(\alpha_i)$ if $x = i$.

Proceeding formally let, in the usual way,

$$V_i(t) = \mathscr{E}\left\{\int_0^t L[x(s), \alpha(x(s))]\, ds \mid x(0) = i\right\},$$

and write $V = (V_1, V_2, \ldots)'$, $L = (L_1, L_2, \ldots)'$. Then V satisfies the differential equation

$$dV(t)/dt = Q(\alpha)V(t) + L(\alpha), \qquad t \geq 0,$$
$$V(0) = 0.$$

On the assumption that the corresponding $x(\cdot)$ process admits a unique ergodic probability measure, with respect to which $L(x, \alpha(x))$ is integrable, we shall have

$$\mathscr{E}\{L[x, \alpha(x)]\} = \lambda 1$$
$$V(t) \sim \lambda 1 t + R, \qquad t \to \infty,$$

for some constant scalar λ and vector R, where $1 = (1, 1, \ldots)'$. Noting that $Q1 = 0$ we obtain by formal substitution

$$\lambda 1 = Q(\alpha)R + L(\alpha).$$

This result is the heuristic basis for Bellman's equation

$$\min_{\alpha}[Q(\alpha)R + L(\alpha)] = \lambda 1. \tag{7.3}$$

In (7.3) the minimization is componentwise; and R, λ are to be determined. The fact that (7.3) provides a sufficient condition for optimality is established by an argument similar to that of Section III.E.

Just as in Section III.E, the existence of a control $\alpha \in \mathscr{A}$, for which the corresponding $x(\cdot)$ process is ergodic and $|\lambda| < \infty$, is a stability (or stabilizability) property. If the state-space $\mathscr{X}$ is finite then it is well known that $x(t)$ is with probability 1 ultimately absorbed in an ergodic subset of $\mathscr{X}$; in this case the existence problem for (7.3) is relatively straightforward. The situation is much less simple if $\mathscr{X}$ is denumerable, but sufficient conditions for stabilizability can be based on inequalities of type $\lambda 1 \geq Q(\alpha)R + L(\alpha)$ (cf. Cox and Miller [8a], Chapter 3, §8).

For detailed discussion of the asymptotic behavior of $V(t, x)$ and existence of R, the reader is referred to White [69a], Romanovskii [59a] and the report by Fife [13a].

The problem of control on a finite time interval has been investigated in detail by Miller [51b].

C. APPROXIMATIONS TO OPTIMAL FEEDBACK CONTROLS

We have seen that the (Bellman) functional equations of dynamic programming which arise in the formulation of stochastic optimization problems are typically of (integro-) partial differential type. In the special cases considered in Sections III–VI, the equations can be reduced by appropriate substitutions to simpler types for which numerical computation is generally feasible, if the state space dimension n is not too large.[5] These special cases apart, computational difficulties are severe. It is hard, and very expensive, to solve Bellman's equation by finite differences if $n > 2$; and a purely numerical solution would require considerable further processing to yield a representation of the control useful for practical purposes. The situation is not helped by adopting a discrete-time model at the start.

[5] Currently the upper limit is $n \sim 20$.

The problem is basically one of adopting a plausible (if approximate) parametric representation of the optimal control. A direct method, sometimes effective, is then to simulate the system and stochastic inputs, and to use a gradient technique in parameter space to seek optimal parameter values. Here the difficulties are well-known: the procedure is slow, is prone to stick at relative (false) minima, and may suffer from poor resolution due to unsuitable parametrization of the family of controls.

A less naive approach is to exploit the functional equation as a guide, but without attempting to solve it "exactly." As an example (Wonham and Cashman [82]) consider the problem of determining optimal stationary feedback control $u = \phi(x)$ for the system

$$dx = Ax\,dt - Bu\,dt + C\,dw$$
$$\mathcal{E}_\phi\{x'Mx + \phi(x)'N\phi(x)\} = \text{minimum},$$

subject to the constraint

$$\max_i |\phi_i(x)| \le 1, \qquad x \in R^n \qquad (7.4)$$

Here the parameter matrices A, B, C, M, N are constant; $w(\cdot)$ is a Wiener process; and the notation $\mathcal{E}_\phi$ has the same meaning as in Section III.E. Proceeding just as in the section cited, we find that Bellman's equation is a quasilinear elliptic equation. If the solution is $V(x)$, the optimal control is given by

$$\phi^0(x) = \text{sat}[(1/2)N^{-1}B'V_x(x)] \qquad (7.5)$$

where $(\text{sat } y)_i = y_i$ or $\text{sgn } y_i$ according as $|y_i| \le 1$ or ≥ 1. Thus the constraint (7.4) implies that the optimal control is nonlinear. One effective computational scheme (for details see Wonham and Cashman [82]) relies on the crude assumption that $V(x) \doteq x'Px$ for suitable $P > 0$ and that a control of form (7.5) can be represented as

$$\phi(x) \doteq k'x$$

for a suitable vector k; k is chosen by statistical linearization (Booton [6]). On this basis the exact (partial differential) equations of approximation in policy space are mimicked by algebraic equations for P and k. "Optimal" values of P and k are obtained in a few seconds by digital computer, and the control implemented is

$$[\phi^0(x)]_{\text{approx}} = \text{sat}(N^{-1}B'Px). \qquad (7.6)$$

Evidence suggests that the control (7.6) is satisfactorily close to the optimal, in terms of the corresponding quadratic cost. This scheme has been used successfully with systems of dynamic order up to 8, for which an "exact" solution of Bellman's equation would be out of the question.

The conclusion is that optimal control theory may suggest useful practical approaches to stochastic, as well as "deterministic," problems of feedback control; but only when combined with insight into the concrete problem at hand.

For further discussion of approximate methods the reader is referred to the books by Pervozvanskii [53] and Aoki [1].

D. OPEN-LOOP CONTROL AND SAMPLED SYSTEMS

The relative merits of open-loop and feedback control were discussed in a general way in Section II.D. In many applications the continuous use of feedback control may require complex instrumentation and actually be unnecessary to meet performance specifications; whereas the opposite extreme of pure open-loop or preprogrammed control would result in excessive error due to disturbances. A practical compromise is to divide the process time interval into segments with pre-assigned end-points or *sampling instants*. The dynamic state $x(\cdot)$ of the system is observed or estimated only at the sampling instants t_ν and control in the interval $(t_\nu, t_{\nu+1})$ is determined from $x(t_\nu)$ on an open-loop basis.

Such an approach is common in the guidance of space vehicles (Jordan [31], Jazwinski [30]). For the method to be effective, the sampling intervals must be short enough so that the variance of the state at the endpoints $t_\nu - 0$ is kept "small," this being interpreted ultimately by consideration of the corrective forces allowed. Approximately "optimal" open-loop controls are usually determined by perturbation techniques which render the problem tractable by ordinary calculus of variations, as in the articles just cited. A more fundamental approach is to treat the stochastic open-loop problem as one of nonlinear programming in function space: see, for example, Kirillova [40], Kozhev-nikov [41], and Van Slyke and Wets [68, 69].

Finally we note that, when cost of observation and computation is high, the intelligent location of sampling points becomes important and the situation as a whole becomes far more complex (see, e.g.,

Krasovskii [44]). We encounter here the problem of matching informa-
tion-handling facilities with control requirements, and for this a general
theory is currently lacking.

E. STOCHASTIC DIFFERENTIAL EQUATIONS AND NONLINEAR FILTERING

The practical appeal of a dynamic filtering theory was pointed out in
Section IV. It is natural to attempt a generalization to the case where
the equations of plant dynamics and channel may be nonlinear; so
that (2.5) is replaced by

$$dx = f(t, x) \, dt + G(t, x) \, dw,$$
$$dy = h(t, x) \, dt + K(t) \, dw. \tag{7.7}$$

A difficulty which arises immediately with (7.7) is that the conditional
distribution $P(t, x)$ of $x(t)$ given $y(s)$, $s \leq t$, usually cannot be specified
as to type and cannot be characterized by a small number of parameters
(e.g., mean and variance) which would serve as sufficient statistics. One
is thus led to the problem of determining the stochastic equation of
evolution for $P(t, x)$ itself or for sets (typically, infinite) of moment
functions

$$m_v(t) = \int \alpha_v(x) P(t, x) \, dx, \qquad (v = 1, 2, \ldots), \tag{7.8}$$

with suitably chosen α_v.

In Section V it was shown that, when (7.8) is linear in x, a large class
of control problems could be solved as a standard dynamic programming problem in the state space of values of a single sufficient statistic
$\hat{x}(t)$. In the present case $\hat{x}$ would be replaced formally by $P(t, x)$ or by
the infinite system $\{m_v\}$; but the optimization problem has not yet
been precisely formulated along these lines and to do so might perhaps
be only of academic interest.

In practice the infinite system (7.8) is replaced by some approximating
finite subsystem, from which an estimate $\hat{x}(t)$ of $x(t)$ is determined in
some "reasonable" way; the problems of control and filtering are thus
forcibly separated at the start. Whether or not much loss of performance
is incurred by separation is an interesting unsolved theoretical question.
The question is fundamental in regard to complex systems with an
adaptive capability, where control signals may play a dual role of
control proper and of "probing" or identification (Feldbaum [12];
Florentin [20]).

For a detailed discussion of the filtering problem the reader is referred to the pioneering paper of Stratonovich [64] and to more recent and rigorous studies by Shiryaev [61] and Kallianpur and Striebel [33]. A comparison of various approximate nonlinear sequential digital filtering schemes is given in reports by Meier [57a] and Athans *et al.* [2]. The area is one of active research.

ACKNOWLEDGMENTS

This article is based on lectures given by the author in the Center for Dynamical Systems, Division of Applied Mathematics, Brown University, in 1966–1967. Most of the article was written during the author's tenure of an NRC Senior Postdoctoral Resident Research Associateship, supported by the National Aeronautics and Space Administration, and pursued in the Office of Control Theory and Application of the NASA Electronics Research Center.

REFERENCES

1. Aoki, M., "Optimization of Stochastic Systems," Academic Press, New York, 1967.
2. Athans, M., Wishner, R. P., and Bertolini, A., Suboptimal state estimation for continuous time nonlinear systems from discrete noisy measurements, *Trans. IEEE* **AC-13** (1968), 504–514.
3. Barrett, J. F., Application of Kolmogorov's equations to randomly disturbed automatic control systems, in "Automatic and Remote Control," Vol. 2, pp. 724–733, Butterworth, London, 1961 (Proc. IFAC, 1960).
4. Bellman, R., "Dynamic Programming," Princeton Univ. Press, Princeton, New Jersey, 1957.
5. Bellman, R., "Adaptive Control Processes, A Guided Tour," Princeton Univ. Press, Princeton, New Jersey, 1961.
6. Booton, R. C., Jr., Nonlinear control systems with random inputs, *Trans. IRE PGCT* **CT-1**, 9–18, 1954.
7. Chuang, K., and Kazda, L. F., A study of nonlinear systems with random inputs, *Trans. AIEE* **78** (1959), 100–105.
8. Clark, J. M. C., The Representation of Non-Linear Stochastic Systems with Applications to Filtering, Ph.D. Thesis, Electrical Engineering Department, Imperial College, London, 1966.
8a. Cox, D. R. and Miller, H. D., "The Theory of Stochastic Processes," Wiley, New York, 1965.
9. Dashevski, M. L., and Lipster, R. Sh., Simulation of stochastic differential equations connected with the "disorder" problem by means of analog computers, *Automat. Remote Control* **27** (1966), 665–673.

10. Doob, J. L., "Stochastic Processes," Wiley, New York, 1953.

11. Dynkin, E. B., "Markov Processes," Academic Press, New York, 1965.

12. Feldbaum, A. A., Dual control theory I-IV, *Automat. Remote Control* **21** (1960), 1240–1249; **21** (1960),1453–1464; **22** (1961), 3–16; **22** (1961), 129–143.

13. Feldbaum, A. A., Problems in the statistical theory of systems of automatic optimization, *in* "Automatic and Remote Control," Vol. 2, pp. 547–555, Butterworth, London, 1961 (Proc. IFAC, 1960).

13a. Fife, D. W., "Optimal Control of Queues with Application to Computer Systems," Tech. Rept. 170, Cooley Electronics Lab., Univ. of Michigan, October, 1965.

14. Fleming, W. H., Some Markovian optimization problems, *J. Math. Mech.* **12** (1963), 131–140.

15. Fleming, W. H., The Cauchy problem for degenerate parabolic equations, *J. Math. Mech.* **13** (1964), 987–1008.

16. Fleming, W. H., Duality and *a priori* estimates in Markovian optimization problems, *J. Math. Anal. Appl.* **16** (1966), 254–279.

17. Fleming, W. H., Optimal control of partially observable diffusions, *SIAM J. Control* **6** (1968), 194–214.

18. Fleming, W. H., and Nisio, N., On the existence of optimal stochastic controls, *J. Math. Mech.* **15** (1966), 777–794.

19. Friedland, B., Thau, F. E., and Sarachik, P. E., Stability problems in randomly excited dynamic systems, Proc. 1966 Joint Automatic Control Conference, Seattle, 1966, pp. 848–861.

20. Florentin, J. J., Optimal, probing, adaptive control of a simple Bayesian system, *J. Electronics Control* **13** (1962), 165–177.

21. Friedman, A., "Partial Differential Equations of Parabolic Type," Prentice-Hall, Englewood Cliffs, New Jersey, 1964.

22. Gikhman, I. I., and Skorokhod, A. V., "Introduction to the Theory of Random Processes" (in Russian), Izd.-vo "Nauka," Moscow, 1965.

23. Girsanov, I. V., Minimax problems in the theory of diffusion problems, (in Russian) *Dokl. Akad. Nauk SSSR* **136** (1961), 761–764.

24. Howard, R. A., "Dynamic Programming and Markov Processes," Wiley, New York, 1960.

25. Howard, R. A., Semi-Markovian decision processes, *Bull. Inst. Internat. Statist.* **40** (1963), 625–652.

26. Il'in, A. M., and Khas'minskii, R. Z., On the equations of Brownian motion, *Theor. Prob. Appl.* **9** (1964), 421–444.

27. Itô, K., Stochastic differential equations in a differentiable manifold, *Nagoya Math. J.* **1** (1950), 35–47.

28. Itô, K., On stochastic differential equations, *Mem. Amer. Math. Soc.* **4**, 1951.

29. Jacobs, O. L. R., and Wonham, W. M., Extremum control in the presence of noise, *J. Electronics Control* **3** (1961), 194–211.

30. Jazwinski, A. H., On optimal stochastic midcourse guidance, *J. Optimization Theory Appl.* **2** (1968), 331–347.

31. Jordan, J. F., Jr., Optimal Stochastic Control Theory Applied to Interplanetary Guidance, Eng. Mechanics Research Lab., The Univ. of Texas, Austin, Texas, EMRL Tech. Rept. 1004, August, 1966.

32. Joseph, P. D., and Tou, J. T., On linear control theory, *Trans. AIEE* **80** (1961), 193–196.
33. Kallianpur, G., and Striebel, C., Stochastic Differential Equations Occurring in the Estimation of Continuous Parameter Stochastic Processes, Tech. Rept. No. 103, Dept. of Statistics, Univ. of Minnesota, September, 1967.
34. Kalman, R. E., A new approach to linear filtering and prediction problems, *J. Basic Eng.* March (1960), 35–45.
35. Kalman, R. E., New Methods in Wiener Filtering Theory. Proc. First Symp. on Eng. Appl. of Random Function Theory and Probability, Wiley, New York, 1963, pp. 270–388.
36. Kalman, R. E., and Bucy, R. S., New results in linear filtering and prediction theory, *J. Basic Eng.*, March (1961), 95–108.
37. Khas'minskii, R. Z., Ergodic properties of recurrent diffusion processes and stabilization of the solution to the Cauchy problem for parabolic equations, *Theor. Prob. Appl.* **5** (1960), 179–196.
38. Khas'minskii, R. Z., A limit theorem for the solutions of differential equations with random right-hand sides, *Theor. Prob. Appl.* **11** (1966), 390–406.
39. Khazen, E. M., Evaluation of the one-dimensional probability densities and moments of a random process in the output of an essentially non-linear system, *Theor. Prob. Appl.* **6** (1961), 117–123.
40. Kirillova, F. M., On the problem of existence of an optimal control for a linear system with random disturbances (in Russian), *Mat. Sb.* **5** (1964), 86–93.
41. Kozhevnikov, Yu. V., The principle of optimality in the mean for discontinuous stochastic systems, *Automat. Remote Control* **27** (1966), 1711–1720.
42. Krasovskii, N. N., On optimal control in the presence of random disturbances, *Appl. Math. Mech.* **24** (1960), 82–102.
43. Krasovskii, N. N., Stabilization of systems in which noise is dependent on the value of the control signal, *Engrg. Cybernetics* (1965), 94–102.
44. Krasovskii, N. N., On optimum control with discrete feedback signals (in Russian), *Differencial'nye Uravnenija* **1** (1965), 1415–1427.
45. Krasovskii, N. N., and Lidskii, E. A., Analytical design of controllers in stochastic systems with velocity-limited controlling action, *Appl. Math. Mech.* **25** (1961), 627–643.
46. Krasovskii, N. N., and Lidskii, E. A., Analytical design of controllers in systems with random attributes, I–III, *Automat. Remote Control* **22** (1961), 1021–1025; **22** (1961), 1141–1146; **22** (1961), 1289–1294.
47. Ladizhenskaya, O. A., Solonnikov, V. A., and Uraltseva, N. N., "Linear and Quasilinear Equations of Parabolic Type " (in Russian), Izd.-vo "Nauka," Moscow, 1967.
48. Langevin, P., Sur la théorie du mouvement brownien, *C.R. Acad. Sci. Paris* **146** (1908), 530–533.
49. Luh, J. Y. S., Optimization of Stochastic Control Processes with Respect to Probability of Entering a Target Manifold, Tech. Rept. EE67-15, School of Elect. Engrg., Purdue Univ., October 1967.
50. Mandl, P., On the control of non-terminating diffusion processes, *Theor. Prob. Appl.* **9** (1964), 591–603.
51. Mandl, P., On the control of a Wiener process for a limited number of switchings, *Theor. Prob. Appl.* **12** (1967), 68–76.

51a. Meier, L., " Combined Optimum Control and Estimation Theory," Contractor Rept. NAS-2-2457, Stanford Research Inst., Menlo Park, California, 1965.

51b. Miller, B. L., Finite state continuous time Markov decision processes with a finite planning horizon, *SIAM J. Control* 6 (1968), 266–280.

52. Newton, G. C., Gould, L. A., and Kaiser, J. F., "Analytical Design of Linear Feedback Controls," Wiley, New York, 1957.

53. Pervozvanskii, A. A., "Random Processes in Nonlinear Control Systems," Academic Press, New York, 1968.

54. Peschon, J., and Larson, R. E., "Analysis of an Intercept System," Final Report, SRI Project 5188, Stanford Research Institute, Menlo Park, California, December, 1965.

55. Pontryagin, L., Andronov, A., and Vitt, A., On the statistical investigation of dynamic systems, (in Russian) *Zh. Eksper. i Teor. Fiz.* 3 (1933), 165–180.

56. Pontryagin, L. S., Boltyanskii, V. G., Gamkrelidze, R. C., and Mishchenko, E. F., "The Mathematical Theory of Optimal Processes," New York, Wiley (Interscience), 1962.

57. Potter, J. E., "A Guidance-Navigation Separation Theorem," M.I.T. Exper. Astronom. Lab. Rept. RE-11, 1964.

58. Prokhorov, Yu. V., Convergence in random processes and limit theorems in probability theory, *Theor. Prob. Appl.* 1 (1956), 157–214.

59. Riordan, J., " Stochastic Service Systems," Wiley, New York, 1962.

59a. Romanovskii, I. V., The existence of an optimal stationary policy in a Markovian decision process *Theor. Prob. Appl.* 10 (1965), 130–133.

60. Rozanov, Yu. A., " Stationary Random Processes," Holden-Day, San Francisco, 1967.

61. Shiryaev, A. N., On stochastic equations in the theory of conditional Markov processes, *Theor. Prob. Appl.* 11 (1966), 179–184.

62. Skorokhod, A. V., Limit theorems for stochastic processes, *Theor. Prob. Appl.* 1 (1956), 261–290.

63. Skorokhod, A. V., "Studies in the Theory of Random Processes," Addison-Wesley, Reading, Massachusetts, 1965.

64. Stratonovich, R. L., Conditional Markov processes, *Theor. Prob. Appl.* 5 (1960), 156–178.

65. Stratonovich, R. L., A new representation for stochastic integrals and equations (in Russian), *Vestn. Moskov. Univ. Ser. I. Mat. Meh.* (1964) 3–12. Eng. Trans. *SIAM J. Control* 4 (1966), 362–371.

66. Striebel, C., Sufficient statistics in the optimum control of stochastic systems, *J. Math. Anal. Appl.* 12 (1965), 576–592.

67. Van Mellaert, L. J., "Inclusion-Probability-Optimal Control," Res. Rpt. PIBMRI 1364-67, Polytechnic Inst., Brooklyn, May, 1967.

68. Van Slyke, R., and Wets, R., Programming under uncertainty and stochastic optimal control, *SIAM J. Control* 4 (1966), 179–193.

69. Van Slyke, R., and Wets, R., " Stochastic Programs in Abstract Spaces," Boeing Sci. Res. Labs. D-182-0672, October, 1967.

69a. White, D. J., Dynamic programming, Markov chains, and the method of successive approximations, *J. Math. Anal. Appl.* 6 (1963), 373–376.

70. Wiener, N., "Extrapolation, Interpolation and Smoothing of Stationary Time Series," Wiley, New York, 1949.

71. Wong, E., and Zakai, M., On the convergence of ordinary integrals to stochastic integrals, *Ann. Math. Statist.* **36** (1965), 1560–1564.
72. Wong, E., and Zakai, M., On the relation between ordinary and stochastic differential equations, *Internat. J. Engrg. Sci.* **3** (1965), 213–229.
73. Wong, E., and Zakai, M., On the relation between ordinary and stochastic differential equations and applications to stochastic problems in control theory, Proc. Third Intnl. Congress, IFAC, London, 1966.
74. Wonham, W. M., "Stochastic Problems in Optimal Control," RIAS Tech. Rept. 63–14, May, 1963.
75. Wonham, W. M., Liapunov criteria for weak stochastic stability, *J. Differential Equations*, **2** (1966), 195–207.
76. Wonham, W. M., A Liapunov method for the estimation of statistical averages, *J. Differential Equations*, **2** (1966), 365–377.
77. Wonham, W. M., Optimal stationary control of a linear system with state-dependent noise, *SIAM J. Control* **5** (1967), 486–500.
78. Wonham, W. M., On pole assignment in multi-input controllable linear systems, *Trans. IEEE* **AC-12** (1967), 660–665.
79. Wonham, W. M., "Lecture Notes on Stochastic Control," Lecture Notes 67-2, Center for Dynamical Systems, Division of Applied Mathematics, Brown University, February, 1967.
80. Wonham, W. M., On the separation theorem of stochastic control, *SIAM J. Control* **6** (1968), 312–326.
81. Wonham, W. M., On a matrix Riccati equation of stochastic control, *SIAM J. Control* **6** (1968), 681–697.
82. Wonham, W. M., and Cashman, W. F., A computational approach to optimal control of stochastic saturating systems, *Internat. J. Control* **10** (1969), 77–98.

AUTHOR INDEX

Numbers in parentheses are reference numbers. Numbers in italics indicate the pages on which complete references are listed.

213

SUBJECT INDEX

A

Admissible control, 150, 156, 176
Algebraic equation(s), 2
  root of, 2
  solution set of, 2
  zero of, 2
Antilattice, 90
Atom, 93
Axiomatic model for quantum mechanics, postulates, 108

B

Bellman's equation, 158, 177, 178, 186, 190, 192, 202, 204–206
Boolean σ-algebra(s), 76
  separable, 78

C

Cauchy random polynomial(s), 17, 18
Characteristic observable, 91
Classical quantum mechanics, axioms of, 69
Condensed distribution of roots, 41–43
Configuration space, 67
Contact transformation(s), 65
  infinitesimal, 66
Containment probability, 201
Control(s)
  admissible, 150, 156, 176
  closed-loop, 153
  feedback, 149, 150, 152, 156, 200, 204
  open-loop, 148, 152, 153, 206
  optimal, 148, 150, 156, 200, 204
  stationary, 154, 205

Coordinate functions, 62
Coordinates, 62

D

Degrees of freedom, 62
Diffusion process(es), 136, 155, 178, 200
  differential generator of, 137
Dynamic state, 146, 147
Dynamic system
  description by stochastic differential equation(s), 143, 144
  state of, 134
Dynamical group, 64, 114, 117, 118
  infinitesimal generator of, 64
Dynamical operator, 118
Dynamical variable, 65

E

Error
  covariance of, 164
  minimum weighted expected squared, 163
Estimate
  linear, 163
  unbiased, 163
External control system, 201

F

Feedback law, 149
Fieller's theorem, 35
Filter equations, 167–173
Fundamental principle of classical mechanics, 63
Fundamental theorem of algebra, 2